ANTICIPATING FUTURE ENVIRONMENTS

ANTICIPATING FUTURE ENVIRONMENTS

CLIMATE CHANGE,
ADAPTIVE RESTORATION,
AND THE COLUMBIA RIVER BASIN

SHANA LEE HIRSCH

UNIVERSITY OF WASHINGTON PRESS | *Seattle*

Anticipating Future Environments was supported by a generous grant from the Tulalip Tribes Charitable Fund, which provides the opportunity for a sustainable and healthy community for all.

www.tulalipcares.org

Copyright © 2020 by the University of Washington Press

Design by Katrina Noble
Composed in Iowan Old Style, typeface designed by John Downer

24 23 22 21 20   5 4 3 2 1

Printed and bound in the United States of America

All rights reserved. No part of this publication may be reproduced or transmitted in any form or by any means, electronic or mechanical, including photocopy, recording, or any information storage or retrieval system, without permission in writing from the publisher.

UNIVERSITY OF WASHINGTON PRESS
uwapress.uw.edu

LIBRARY OF CONGRESS CATALOGING-IN-PUBLICATION DATA
Names: Hirsch, Shana Lee, author.
Title: Anticipating future environments : climate change, adaptive restoration, and the Columbia River Basin / Shana Lee Hirsch.
Description: Seattle : University of Washington Press, [2020] | Includes bibliographical references and index.
Identifiers: LCCN 2019053533 (print) | LCCN 2019053534 (ebook) | ISBN 9780295747491 (hardcover) | ISBN 9780295747293 (paperback) | ISBN 9780295747484 (ebook)
Subjects: LCSH: Watershed restoration—Columbia River Watershed. | Restoration ecology—Columbia River Watershed. | Wildlife resources—Climatic factors.
Classification: LCC QH104.5.C64 H57 2020 (print) | LCC QH104.5.C64 (ebook) | DDC 333.73/15309711—dc23
LC record available at https://lccn.loc.gov/2019053533
LC ebook record available at https://lccn.loc.gov/2019053534

COVER DESIGN: Katrina Noble
COVER PHOTOGRAPH: South Fork of the Snake River, Idaho. The Snake River ultimately flows into the Columbia River in southeast Washington. Courtesy of the Bureau of Land Management.

The paper used in this publication is acid free and meets the minimum requirements of American National Standard for Information Sciences—Permanence of Paper for Printed Library Materials, ANSI Z39.48-1984.∞

To my Dad, who has taught me how to do so many things.
He might not believe in climate change, but he has always believed in me.

Knowledge is high in the head, but the salmon of wisdom swims deep.

—NEIL GUNN

CONTENTS

Acknowledgments ix
List of Abbreviations xi

Introduction: No Small Task 3
  1. A Science for the Columbia River Basin 38
  2. River Restoration in the Columbia River Basin 62
  3. The Work of Restoration in a Changing Climate 88
  4. Emergence: Making Room for the River to Breathe 104
  5. Acclimation: Rethinking Monitoring for the Future 123
  6. Anticipation: Imagining a Future 146
Conclusions: Environmental Imaginaries and River Futures 166

Notes 185
References 187
Index 205

ACKNOWLEDGMENTS

This book would not have been possible without the support and encouragement of many others. At the top of this long list is my mentor, Jerrold Long. Thank you for encouraging this oftentimes messy interdisciplinary work. Your thoughtful and honest advice, support, and perseverance in helping me clarify my arguments, my ideas, and my writing has made this book what it is. Thank you to Jennifer Ladino, Adam Sowards, and Brian Kennedy for their help and support in developing this project and their advice in seeing it through. And, especially, thank you to Barbara Cosens, for her mentorship and inspiration in pursuing adaptive governance scholarship. Thank you also to Steve Daley-Laursen for presenting me with so many opportunities to engage with transdisciplinary work and to develop my leadership skills. Thanks also to other mentors who have inspired and challenged me along the way, including Ben Gardner and David Ribes.

Funding for this research was provided by the National Science Foundation (NSF) through its Integrative Graduate Education and Research Traineeship (IGERT) program (no. 1249400). The opportunities for collaborative work provided to me through IGERT laid a foundation for this book, and I thank Tim Link for his leadership of the program at the University of Idaho. Additional funding for fieldwork was provided by another NSF grant (no. 1655884). Funding was also provided in the form of a fellowship from the United States Geological Survey's Northwest Climate Science Center.

Further, thank you to all of my colleagues in the IGERT program. Our conversations inspired me and provided me with the motivation and ideas for pursuing this project. I owe a similar debt to my colleagues and friends at universities near and far, and especially to the women in my science and technology studies reading and professional development group: Lauren Drakopulos, Sarah Inman, and Anissa Tanweer. I am grateful to the staff at the University of Washington Press and to the two anonymous readers who also provided many helpful and critical insights and encouraged me to push the ideas in this book further.

Deep gratitude to my family and my friends, and especially to Justin Hirsch, who not only made this work physically and mentally possible but endlessly encouraged me to pursue it. A special thank you to my family: Victoria Hunter, Dimitri Hunter, David Hunter, Susan Hunter, Glenn Hirsch, Sue Dicker, Danielle Hirsch, Girin Guha, and Bran, who have also supported me through the project from the beginning.

Finally, this is an ethnographic work and it would not have been possible without the many individuals throughout the Columbia River Basin who gave up precious hours of their workdays (or days off), inviting me into their offices and field sites to share their hopes, dreams, and fears about the restoration of salmon to the Columbia River Basin. I feel humbled and inspired to have had the opportunity to speak to such a diverse group of dedicated people, many of whom have devoted decades of their lives and their entire careers to the goal of salmon recovery. I hope that their dedication and effort can be felt in their words and their work depicted in this book.

Parts of this book were previously published in different forms in other venues. Parts of chapter 2 were published in the journal *Environmental Values* (Hirsch & Long, 2018). Sections of the book, and especially chapters 3 and 6, were published in *Environment and Planning E: Nature and Space* (Hirsch, 2019), and other sections throughout the book appear in different form in *Science, Technology, & Human Values* (Hirsch & Long, 2020).

ABBREVIATIONS

AEM	Action Effectiveness Monitoring of Tributary Habitat Improvement
AMIP	adaptive management implementation plan
BDA	beaver dam analog
BiOp	biological opinion
BoR	Bureau of Reclamation
BPA	Bonneville Power Administration
CBD	Convention for Biological Diversity
CHaMP	Columbia Habitat Monitoring Program
CRITFC	Columbia River Inter-Tribal Fish Commission
ESA	Endangered Species Act
FCRPS	Federal Columbia River Power System
IMW	intensively managed watershed
ISEMP	Integrated Status and Effectiveness Monitoring Program
ISRP	Independent Scientific Review Panel
ISTM	Integrated Status and Trend Monitoring Project
LWD	large woody debris
NEPA	National Environmental Policy Act
NOAA Fisheries	National Oceanic and Atmospheric Administration Marine Fisheries Service
NW Council	Northwest Power and Conservation Council
NWI	Northwest Watershed Institute
PIT	passive integrated responder

PNAMP	Pacific Northwest Aquatic Monitoring Partnership
RPA	reasonable and prudent alternative
STS	science and technology studies
UNEP	United Nations Environment Program
USFS	United States Forest Service

ANTICIPATING FUTURE ENVIRONMENTS

Introduction

No Small Task

Until I physically attempted to travel the length and breadth of the Columbia River and its tributaries, I had no way to conceptualize its vastness. Depending on where you are headed, driving across the Columbia River Basin on an interstate highway takes a day or more. If you are traveling into the mountain headwaters, you should be prepared to drive for several more hours on gravel roads. But the Columbia River Basin is not only vast, it is also infinitely diverse. The ecology shifts mile by mile; one hour you might be crossing a high plateau—driving through sage-scrublands, dodging tumbleweeds, or passing through irrigated farmlands, orchards, and fields of fragrant mint. An hour later, altitude and distance will shift the ecology, and as you pass over and through one after another of the many major mountain ranges, the high alpine forests will dissolve into ponderosa pine. These ecotones of shifting vegetation blur down into valleys of grasslands dotted with distant black cattle, cottonwoods along the creeks, and swarms of hatching insects and humming crickets. The seasons also transform the landscape. The layering of ecosystems and time mean that no journey through the basin is ever the same: along the Columbia, deep snow pushes elk to the lowlands in winter, and rattlesnakes sun themselves in summer. The sandhill cranes and white pelicans make their way south through the dotted wetlands in the spring, greeted along the way by the Othello Sandhill Crane Festival. On my journeys across the Columbia River Basin, I saw pink skies, silver coyotes, antelope, badgers, and countless pairs of ravens. And I was constantly reminded that the river gives us all of these things. Without the Columbia and the water that flows into it, none of what I have just described would exist.

In researching this book over the course of four years, I added 50,000 miles to my odometer. While this requires some dedication on the part of the driver, the salmon that have migrated back to the Columbia's remote headwater streams make my road trip seem trivial in comparison. Some of these fish have traveled hundreds of miles and passed over eight major dams—in each direction—to return to their spawning grounds. Whether they are hatchery or wild fish, they connect the ocean to remote mountain spawning grounds by bringing back marine nutrients in their flesh.

The complexity of the lives of these fish and the environmental and social makeup of the entire Columbia River Basin often seems overwhelming. The large scale of the basin, the diverse ecology and complexity of salmon life cycles, the devastating legacy of colonialism, and the development of one of the boldest engineering projects in the world, are all written in the landscape and the lives of those who live here. One way—and by no means the only way—that people throughout the basin have been working to understand this complexity is by employing the tools of science. And now, a vast network of ecological restorationists are attempting to reverse the environmental degradation that has brought many of the Columbia River's salmon species close to extinction. Today, restoration of salmon habitat has more political and economic support than it has ever had; and over the past few decades, a large economy, or even what some refer to as an "industry" has emerged to drive this restoration effort forward. Parallel to this, an epistemic community (Haas, 1992) of restorationists—a diverse group of people involved in understanding, planning, and implementing ecological restoration of salmon habitat—has grown to support the work of ecological restoration through ecological monitoring and scientific application of environmental management tools. Although this restoration project is complex in its own right, another layer of complexity has been introduced: climate change.

Like most people involved in restoration that I spoke to as I traveled throughout the Columbia River Basin (and during the course of my research I talked to many people), the restorationist I was on my way this day to meet agreed: climate change has arrived. He described what it was like to be an ecological restorationist living through it:

> Back in 2015, that super-hot summer was destructive. It was wild. It was so hot for so long. . . . The whole main channel was 80 degrees or more. And there would just be these clumps of fish trying to survive where a seep of groundwater was coming out. You could *see* that. You could *see* adult sockeye dying . . . [there were] mass deaths of adult sockeye because they were just too stressed. . . . So you can kind of see how things might end up looking, and it is frightening. . . . That was tough. Trees died. There were fires.

I was meeting with a restoration practitioner who has been doing fieldwork and project management in the Upper Columbia River Basin for over a decade. Driving down the highway that follows the south bank of the Columbia, you can easily miss his field office, where he works for a tribal fisheries department. The office itself is a nondescript porta-cabin wedged between the river and the highway; but from this lowly hub, he and his colleagues working for the tribe have helped shape restoration in the region. As a project manager and restorationist, he has been witness to changes in the environment. He and his colleagues have seen the way that shifts in climate have impacted their own work to restore salmon populations and the habitat that salmon rely on. And although when I spoke to him over a year had passed, the summer of 2015 was still fresh on his mind:

> There were huge fires and my staff in the Methow were on evacuation order for about a month straight. It was really scary. It was a really tough summer. We were all like, "Can we just get to fall? Because this is really brutal."

Unlike some places in the world, where climate change may be having more subtle effects (at least at the time of the writing of this book), change can already be seen and felt in the Columbia River Basin. The winter of 2014–2015 brought a "snow drought" to the—usually white—peaks of the

Cascades and the Northern Rockies. The first six months of 2015 were the warmest ever recorded, and the wildfire season was the most severe the region had experienced in modern history (Blunden & Arndt, 2015; National Interagency Fire Center, 2017; Vano, Nijssen, & Lettenmaier, 2015). "Climate change is something that we have been living through, essentially for the last 50 years," one land manager observed, "you can already see these trends in the weather stations and at the flow-gauging stations."

Scientists corroborate the changing climate that Pacific Northwest residents are witness to. Across the Columbia River Basin, rivers and streams are warmer than they have ever been before (Dalton, Mote, & Snover, 2013; Mote et al., 2016). At higher elevations, most watersheds are experiencing a shift in precipitation from snow to rain, raising water temperatures to levels dangerous to salmon (Mantua, Tohver, & Hamlet, 2010). The changes in timing and amount of snowfall in the mountains are in turn affecting the hydrograph of the rivers and impacting fish migration, spawning, and rearing throughout the year (Rieman & Isaak, 2010). In 2015, hundreds of thousands of salmon perished in a massive die-off due to low river flows and fatally high water temperatures (NW Council, 2015).

To most people in the basin who rely on natural resources in some way, environmental change is becoming increasingly apparent. This includes irrigators, ranchers, and farmers; dam operators, fishermen, and municipalities—but it also includes ecological restorationists. They are working to restore endangered salmon habitat and are being forced to adapt their scientific work and management practices in order to meet these changing conditions. There is a strong consensus that these environmental changes are adding a new layer of complexity and uncertainty to ecological restoration of salmon habitat (Beechie et al., 2013). Yet, while policy makers, scientists, and engineers all recognize that ecological restoration is critical for maintaining biodiversity and mitigating the impacts from climate change, it is still fundamentally unclear whether—and how—salmon can be recovered and their habitat restored to the Columbia River Basin.

For many people in the Pacific Northwest, 2015 was a wake-up call. The winter of 2014–2015 brought record-low snowfall to the Cascades and Rockies. The ensuing high temperatures, drought, and low streamflows

Columbia River Basin salmon migration. (By Marley Slade, adapted from Northwest Power and Conservation Council)

in the spring and summer of 2015 brought environmental devastation—especially to salmon runs. For many in the restoration community, 2015 was a benchmark for how bad things *could* be in the future, and for many it represented an oracle of sorts. Globally, 2015 was the warmest year on record, with 2014 already breaking the previous record (Blunden & Arndt, 2015). In the Pacific Northwest, the warm temperatures contributed to record low snowpack. In many areas, this equated to *zero* snowpack, a drastic decline in spring runoff, and drought conditions throughout the region (Mao, Nijssen, & Lettenmaier, 2015). On top of this, a hot, dry summer increased fire intensity and frequency throughout the basin (Vano, Nijssen, & Lettenmaier, 2015). These high temperatures were deadly to salmon. As the fish made their way up the main stem in early summer, record-high water temperatures exceeded survivable levels, and mass die-offs ensued (NOAA Fisheries, 2015). The Northwest Power and Conservation Council (NW Council) estimates that around 250,000 sockeye died in the Columbia River and its tributaries as a result of the warm water. This included well over half of some runs (NW Council, 2015). Snake River sockeye were particularly hard hit, with only a handful returning to their headwater streams to spawn.

Unfortunately for salmon and other organisms that rely on cold water and dependable stream flows, according to climate models the river of the future does indeed look like the river of 2015. Winters are predicted to be warmer and wetter, and summers will be dryer and hotter, resulting in low snowpacks, decreased stream flows, and increased water temperatures—all detrimental to salmonid survival (Nolin et al., 2012). These changes also introduce more stochasticity, or random fluctuations, to the river system, as precipitation events are expected to become more intense and rain-on-snow events produce a more extreme hydrograph (Crozier, 2016; Vano, Nijssen, & Lettenmaier, 2015). In addition to increasing the mortality of migrating fish and smolts, the juvenile salmon, these changes could be potentially devastating to salmon populations as a whole: shifting the range of salmon species northward, reducing genetic diversity, as well as altering the timing of salmonid runs. All of these changes are expected to accelerate in the Columbia River Basin over the next few decades (Roberts et al., 2015). Shifts in the timing and intensity of runoff will not only affect fish;

they will also change the seasonality of hydropower production—a key consideration for many stakeholders in the region, and a key source of restoration funding itself (Kao et al., 2015).

Professionals working hard to restore local ecosystems throughout the basin are trying to grasp, understand, and adjust to these changes. My interviewee is just one of them:

> 2015 kind of gave us a little bit of a reality check: Oh, wow! . . . It was a really tough summer. . . . It was stressful. It was like, "If this is what it is going to be like, it is not going to be fun." Stream temperatures were just totally lethal. Totally lethal.

Many people are now using the 2015 season as a new benchmark—both an environmental and a legal baseline. And, those who lived through that year now talk about having had a vision into the environmental future of the river basin.

Restorationists are adapting their work to meet the challenges of climate change in different ways; but for many, the year of 2015 represented a key test of these adaptations. Some practical changes include shifts in their work patterns and to their restoration design practices, such as switching to alternate plant species that might tolerate future conditions. For instance, one restoration manager who has worked in the central basin for two decades recalled her experience of 2015:

> We had done a restoration project that we finished in late 2015, and then we were planting native species over the winter. Normally that is fine, and the roots are fine. But 2015 was extremely long and dry and hot in the summer so we ended up having to water our plants for the first time ever. We were like, "well, at this point maybe we should start thinking about more drought tolerant species." Because one of the predictions is that precipitation is going to fall as rain and it is going to fall west of the Cascades and then you'll have a wetter winter but a dryer longer summer. That is what they are predicting. So, 2015 might be what you are considering the new normal when it comes to the future.

This is just one example of the kinds of adjustments to restoration planning and work practices that are taking place throughout the basin. People are having to make incremental changes as they deal with climate change in their daily work. But these adaptations also extend to more fundamental and epistemic changes, including rethinking the focus of restoration ecology itself in order to anticipate the future. These changes are being seen, felt, and acted upon in the Columbia River Basin, and they are—or soon will be—recognized and acted on globally, as climate change occurs and environmental changes become widespread and apparent.

While environmental change is an inevitable challenge that all fields of ecological science and conservation must address, the increasingly apparent environmental shifts in the Columbia River Basin have not only frustrated but also emboldened the restoration community. And it has led to public support. For example, as climate change becomes more detectable to the average person, the public is also becoming more aware of the importance of habitat restoration. Ecological restoration can be used to mitigate the effects of climate change, including warming waters and increasingly dramatic swings in the hydrograph, by reconnecting cooler groundwater, increasing shade and habitat diversity, and strengthening ecological and physical responses to change (Beechie et al., 2012). For restorationists, this is a time to reconsider goals, rethink current tactics, and reimagine a future, climate-changed river. As one restorationist that I met at a conference put it: "Everything has changed. . . . I personally believe that going back is not a realistic goal."

While the restorationists quoted throughout this book are not only concerned about, but intimately aware of, the environmental changes facing their work, they are nevertheless moving ahead with their job to restore salmon habitat. They head out into the field to monitor change or implement their projects; they work in their offices to develop a plan; they meet at conferences to discuss their ideas and their worries for the future. Within these activities, we see the ways that individuals and their communities are adapting their practices to deal with climate change: making incremental steps that shift a broader context, and taking individual actions that help create a river for the future. Science is a form of work—and within the larger institutional systems that drive the restoration of the basin lie the daily practices of scientists and practitioners in the field or in the lab as they go about their work modeling ecosystems, monitoring streams and rivers, and prioritizing and planning for a climate-changed future.

Scholars have highlighted the science–policy interface as a problematic area for effective response to climate change, and they have called for better communication and understanding at this nexus to support more sustainable transitions (Cash et al., 2003). Although many questions need to be answered in order to address pressing problems in environmental and natural resource management—and there is no doubt that scientists play a pivotal role in finding solutions—these solutions can often be more a matter of paying careful attention to politics than simply providing more data (Bocking, 2004).

This book challenges the idea that more, better, or clearer science will solve environmental management problems. Instead it urges us to look for the basis of these problems in different locations, especially in the production of knowledge itself. To address complex environmental issues, we need to understand how knowledge about environmental issues is produced, exchanged, and used by society. This requires paying attention to what goes on as science is practiced, understanding how certain institutions and infrastructures might help facilitate this practice, and imagining what kinds of epistemic cultures and virtues from might lead to the emergence of these practices.

The transformation—or adaptation—of the science of ecological restoration will directly influence the ways in which natural resource managers and scientists respond to climate change, and consequently, that adaptation will shape the future of the Columbia River itself. As I will demonstrate, this is due to the ways in which epistemic communities and legal regimes such as the Endangered Species Act (ESA) co-produce knowledge. Yet, we also need to consider how environmental and sociotechnical imaginaries—which I discuss more fully in the final chapter—influence the ways that nature is managed and known, facilitating particular material futures for the river and for salmon. These futures imaginaries shape the river environment in particular ways—as people manage the Columbia for particular purposes. Although this influence can be seen in the ways that the Columbia River Basin was developed to meet a particular "sociotechnical imaginary," of a highly regulated, power-producing river, it is also particularly important to recognize how a climate-altered nature comes to be valued and imagined in different ways. The ways in which knowledge itself is produced and how these communities of scientists are able to adapt to environmental change are therefore important social dynamics to understand. This book, therefore, explores how a scientific field—ecological restoration—adapts to climate change.

Waiting for certainty from science is not only impossible, but infeasible. In an environmental context, delays in decisions and action can devastate biodiversity. At conferences and in interviews, restorationists I talked to were frank: in some areas, salmon runs are so close to extinction that one or two hot summers could wipe out all of the genetic diversity within a region. If water volumes are not increased and temperatures decreased *now*, it may be "too late" to restore some populations. Although it is often seen as a taboo discussion, "triage" is already being talked about at regional conferences and by high-level policy-makers. These are some of the reasons that drove me to do this work: to help understand what is happening *right now* in restoration science in the region, and what might be done to enable both environmental managers and scientists to be more effective when dealing with climate change in the Columbia River Basin, as well as globally.

WHY CONSIDER ADAPTATION IN SCIENCE?

At its most basic, climate adaptation is an "adjustment in natural or human systems in response to actual or expected climatic stimuli or their effects, which moderates harm or exploits beneficial opportunities" (Intergovernmental Panel on Climate Change, 2007). In terms of normative goals for management or society, adaptation is the ability of a system to recover from or adjust to a disturbance in a desired way (Folke et al., 2005; Nelson, Adger, & Brown, 2007). Yet natural resource crises often transcend science, resulting from much larger political and economic forces than those tasked with managing them may have control over or even understand. In these situations, the value of nature, the role of expertise, and the implications of uncertainty and ignorance influence how management scenarios play out in practice (Bocking, 2004). Science and politics cannot be separated, and this is especially evident in the Columbia River Basin, where science has been increasingly fought over in courts for the past several decades (Blumm & Paulsen, 2012; Doremus & Tarlock, 2005). Yet, these dynamics are not isolated to the case of the Columbia River Basin and the restoration work there. As climate change, or indeed other social or technical changes take place, scientists must adapt their practices to deal with these changes.

In advocating for adaptation in socio-ecological systems, authors often call for increasing knowledge and data collection in order to enhance system sustainability. For example, Elinor Ostrom's (2009) framework for measuring the sustainability of socio-ecological systems highlights the importance of increasing "predictability of system dynamics" so that decision-makers can better estimate potential outcomes. While the need to increase knowledge and the importance of science are frequently discussed, scientific practice is also touched upon in socio-ecological systems literature in some other, more problematic, ways. For instance, integrating social and ecological knowledge is a main goal of socio-ecological systems research, yet this interdisciplinary integration has been fraught with difficulty.

Theoretical models and predictions have often failed to incorporate social complexity and cultural, political, and economic subtleties often go undiscussed (Ostrom, 2009). To fill this gap, Ostrom (2009) and others have, again, called for more detailed work to be done in order to understand how adaptive institutions can be facilitated, and particularly how "knowledge, deliberation, and learning affect institutional change" (Koontz et al., 2015, p. 147). While these are important factors to consider in facilitating adaptation, it is also important to understand how science, and scientists themselves, can be adaptive in order to support these broader goals of socio-ecological systems adaptation. How, then, can we conceptualize the changes that environmental mangers and scientists—and especially ecological restorationists—are faced with? What are some of the ways that people have discussed these very immanent changes in management and scientific practices that are occurring as they cope with climate change?

ECOLOGICAL RESTORATION AS AN ADAPTING
EPISTEMIC COMMUNITY

The practices that make up the scientific work of ecological restoration in the Columbia River Basin are indeed diverse. While some of these practices involve what might commonly come to mind when we think of laboratory or fieldwork—scientists taking measurements, for instance—scientific practices are also on view at conferences such as River Restoration Northwest, where people present their ideas and different cultures are enacted, negotiated, and performed (Hall, 1997). Throughout the region, there is a vast network of people in the field, in the lab, and in the office. They are ecologists, fisheries biologists, hydrologists, engineers, forestry workers, planners, policy-makers, bulldozer operators, summer interns, community volunteers, and "fish wranglers." They are busy measuring, counting, modeling, monitoring, designing, planning, tagging fish, moving dirt, planting trees, and writing policy. All of this work is, to some extent, a form of scientific "practice" in the sense that as it is done it "makes" and "remakes" the field of ecological restoration and restoration's objects of study. The actors in this diverse epistemic community are each in their own ways practicing and performing ecological restoration.

Ecological restoration—and the conservation sciences more broadly—are applied sciences. This means that restorationists are on the "front line" when dealing with and adapting to climate change, interacting with the environment as they measure, monitor, and alter it. As I will portray throughout this book, scientific practice in the "epistemic community" of ecological restoration takes on many forms. Although the concept of epistemic communities originally hails from international relations, conceptualizing ecological restoration as constituting an epistemic community is useful because it is a way of understanding how this diverse set of actors can shape the environment through scientific work, and particularly how ideas play a role in these outcomes (Haas, 2008).

An epistemic community is "a network of professionals with recognized expertise and competence in a particular domain and an authoritative claim to policy-relevant knowledge with that domain or issue-area" (Haas, 1992, p. 3). Importantly, even as individual practices or approaches to solving a problem differ within the epistemic community, community members are still bound together by the fact that they are a network working on a particular environmental issue and making policy-relevant claims related to that issue (Haas, 2008). The case in this book focuses on the restoration community: in particular, those working to restore the Columbia River Basin, using practices that are considered restoration, and being consulted as experts on restoration when needed by policy-makers. Yet, although this book highlights adaptation within a particular epistemic community of ecological restorationists, the concept of epistemic communities is applicable to those working on environmental issues, and even technological change and innovation, throughout the world. The lessons learned about adaptation in epistemic communities from this case are internationally relevant, because they help us understand how scientific concepts and ideas become relevant to policy, as well as how they shift and adapt to changing environmental or social conditions.

One way to understand how these individuals are dealing with climate change is to look at how they are shifting their practices—simply: what are restorationists doing differently in light of climate change? Focusing, in this way, on how current climatic change is influencing restoration work should not be mistaken for assuming that the scientific effort to understand, manage, and restore habitats has ever been straightforward. Defining seemingly simple concepts such as "natural" or "desirable" ecological states has always proven to be problematic for the fields of ecology and natural resource management. Coming to a consensus on the best way to value and regulate nature, while maintaining multiple visions for managing the environment, is inherently difficult. The meaning of restoration, at its core, is suffused with the idea that it is desirable to return an ecosystem to a historic state of some form; yet restorationists have increasingly been questioning whether this "historic fidelity" is even possible (Hobbs, Higgs, & Hall, 2013; Light, Thompson, & Higgs, 2013).

The uncertainty introduced by climate change challenges restorationists who are working to measure and identify ecological thresholds and create management goals (Hobbs, Higgs, & Hall, 2013; Suding & Leger, 2012). Common scientific and management tools such as ecosystem equilibrium models and historic range of variability are no longer reasonable metrics for success, because historic conditions no longer exist (Seastedt, Hobbs, & Suding, 2008). In response to these challenges, adaptive management and planning are often called for, yet it is unclear exactly what this recommendation entails in terms of scientific practice or management actions (Keith et al., 2011). Restoration ecologists have also pointed out that it is difficult to manage for change when many of the practices and policies in place today were developed based on assumptions of a stable climate (West et al., 2009). And, when combining both of these epistemic and policy issues, the urgent desire for solutions within the restoration community is often coupled with a sense that there is not enough time to employ the scientific method. Salmon survival often depends upon making decisions based on scientific assessments of restoration effectiveness; yet many of the people that I spoke to across the basin worried that there is not enough time to establish rigorous monitoring programs and, consequently, rigorous understandings of restoration effects. In the field of ecological restoration, anticipating the future is becoming an increasing preoccupation.

To deal with this conceptual shift, new "paradigms" have been called for (Choi, 2007). New theoretical concepts have been introduced, such as "hybrid ecosystems," which recognize the need to orient toward a future ecological state (Choi, 2004; Hobbs, Higgs, & Harris, 2009), or "novel ecosystems" that contain species combinations that have not previously occurred (Seastedt, Hobbs, & Suding, 2008). Assisted migration—moving species outside of their home range in order to help mitigate loss of biodiversity—may even be necessary (Corlett, 2016; Dunwiddie et al., 2009; Hobbs, Higgs, & Harris, 2009). The idea of protected areas for conservation worldwide has even been called into question as these tie particular species and ecosystems to places that may no longer support them (Hansen et al., 2010; Heller & Zavaleta, 2009). The effects of climate change have also led some key voices in the field to claim that traditional restoration goals are now "unachievable" (Hobbs, Higgs, & Hall, 2013; Zedler, Doherty, & Miller, 2012). This may, indeed, be the case for some Columbia River salmonids, as water temperatures and flows shift drastically, making it difficult for them to survive in much of their historic range (Mantua, Tohver, & Hamlet, 2010).

While theoretical advances in the field of ecological restoration have been crucial in coming to terms with the fact that there will likely be a "no-analog" future (Williams & Jackson, 2007), exactly how these futures are negotiated by ecologists and practitioners is not well understood. Even though climate change raises fundamental theoretical and practical issues for riparian and in-stream habitat restoration in the region, many restoration practitioners still worry that future climate change scenarios are rarely incorporated into restoration planning. Instead, restorationists working in the field often make decisions on a day-to-day basis, experimenting with and altering the river landscape as they attempt to restore salmon habitat, manage for future climatic change, and deal with high levels of scientific uncertainty and indeterminacy. This book asks how the field of ecological restoration is tackling these issues, and how individual scientists are creating practical solutions as they adapt to climate change. How does a field that has been so focused on looking to the past for guidance shift to anticipate the future?[1] And, more broadly, how does a scientific field *itself* adapt to climate change? Ecological restoration in the Columbia River Basin provides an intriguing case for seeking some answers to these questions. But further, these answers are important because climate change is affecting not only restorationists in the Pacific Northwest of the United States, but also environmental scientists and managers globally.

THINKING THROUGH ADAPTATION

Thinking through the adaptive practices of scientists requires framing of the concept of adaptation itself. As discussed earlier, one place to look for the theory of adaptation is the growing body of literature on adaptation and resilience in socio-ecological systems. This literature engages with theories of change, and often calls upon knowledge production to help facilitate those changes by providing data, monitoring, and experimental design (Walters & Holling, 1990). This provides a starting point for conceptualizing change in environmental management and the science that supports it. The conceptual framework for adaptation and resilience in socio-ecological systems draws on ecological concepts and especially on conceptual theory developed by ecologists such as Gunderson and Holling (2002), along with others. The socio-ecological systems framework conceptualizes the complex amalgam of human and natural forces that are in constant interaction. Socio-ecological systems are multilevel systems, with subsystems of separate parts and variables which exist and act within different scales. All of these systems and parts may be understood on their own, but because they interact with each other, proponents of the socio-ecological systems concept believe that it is better to look at them in a holistic way (Ostrom, 2009). By viewing systems as having multiple scales and parts that interact, we are more likely to not only notice but also understand *how* a change in one variable will affect changes in the larger socio-ecological systems (Ostrom, 2009).

Theories of Change in Socio-ecological Systems

In some fields, this systems perspective has transformed the way that natural resource managers view their work. It has especially influenced the way that ecological systems are modeled in that it provides a way for managers to conceptualize the interactions within complex systems (Taylor, 2005). Socio-ecological systems literature also emphasizes the role of scientific knowledge, data, and modeling in facilitating more adaptive environmental management; and work on adaptive environmental management often discusses the need for knowledge to support this adaptation. Yet, the details of this dynamic are unclear. Therefore, this book asks: what kinds of knowledge would be adaptive, and what would the knowledge infrastructures, institutions, and organizations that facilitate this adaptation look like? And, especially, what kinds of practices would signify that this adaptation is occurring?

Adaptive Governance and Adaptive Law

Recent theoretical work on adaptive governance and adaptive law also provide some ways to understand change in socio-ecological systems. Adaptive governance is a framework for governance that takes change and uncertainty in socio-ecological systems into account (Chaffin & Gunderson, 2016; Gunderson & Light, 2006). It proposes a way to manage ecosystems through collaborative and iterative planning based on institutional learning (Olsson et al., 2006), and is a way to deal with uncertainty by allowing for adaptive management. The theory states that if institutions are flexible and dynamic, as opposed to static, they have an increased "adaptive capacity," allowing socio-ecological systems to "reconfigure themselves" without losing their fundamental functions when confronted with change (Folke et al., 2005). Proponents of adaptive governance, therefore, advocate for organizational learning and cross-scale linkages that are more iterative and therefore enable flexibility and innovation that can address mismatches of scale (Garmestani, Allen, & Benson, 2013). Building adaptive capacity requires bringing ecological knowledge to bear when deciding how to govern socio-ecological systems and, in doing so, it can help create "good" governance systems that are democratic, legitimate, and facilitate equity and justice (Cosens, 2010, 2013).

Recent work on adaptive law has built on theories of adaptive governance, conceptualizing and identifying ways that legal institutions can either foster or hinder adaptation and resilience to climate and other social and ecological changes (Cosens, 2010; Cosens et al., 2017; Cosens et al., 2014; Garmestani, Allen, & Benson, 2013). Many environmental laws were developed in the 1960s and 1970s, when a colloquial understanding of ecology hinged on a "balance of nature" that could be predicted and managed accordingly (Garmestani, Allen, & Benson, 2013; Gunderson, 2013). Indeed, part of the inspiration for this work on adaptive law was to better understand how environmental laws like the ESA, which are often characterized as "inflexible," can adapt to new and emerging issues like climate change (Gosnell et al., 2017). Adaptive law, therefore, considers how the adaptive capacity, processes, and structures within legal institutions themselves can adapt while still maintaining legitimacy (Cosens et al., 2017; Cosens et al., 2014).

Adaptive Management

Adaptive environmental management, first proposed by Walters and Hilborn (1976) and then Holling (1978) also recognizes and even embraces the uncertainty and unpredictability of ecosystems, advancing the idea that, in order to be able to adjust to this uncertainty, managers should iteratively revisit and retest assumptions (Walters & Holling, 1990). Adaptive management is conceptualized as a cycle of steps, including: defining management goals, developing alternative strategies to achieve them, and implementing these strategies in order to compare them—mimicking an experimental process. Throughout these steps, new knowledge is gained about which strategies will meet the management goals, and strategies can then be modified in an iterative way, starting again from the beginning of the cycle. Indeed, adaptive management is sometimes framed as a large-scale experiment used to test hypotheses about ecological models (Lee, 1993; Walters & Holling, 1990), and a kind of "adaptive restoration" has been framed in a similar way (Zedler & Callaway, 2003). By using multiple—or shifting—strategies, adaptive management is able to cope with uncertainty by "spreading the risk": if one of the strategies (or hypotheses) fails, the idea is that others likely will succeed (Keith et al., 2011). Therefore, although uncertainty is recognized as inherent, it is "managed" by incremental increases in relative knowledge and shifting actions as this new knowledge is gained (Walters & Holling, 1990). The aim of adaptive management is to ensure that management decisions are appropriate and continually adjusting to ecological, social, and technoscientific change.

Adaptive management has been practiced by natural resource management agencies for decades. As outlined above, the original framework for adaptive management involved implementing multiple actions at once and determining their success through a comparative experiment (Walters & Holling, 1990). But, in practice, there have been almost no examples in which this strict experimental form of adaptive management has actually been applied to a management issue (Keith et al., 2011). Even when multiple actions are taken, they are rarely separately monitored and then rigorously compared in order to determine relative success. Instead, adaptive management is most often conceptualized as a single, ongoing experiment (Keith et al., 2011). Some have criticized this kind of adaptive management as being more about trial and error (Duncan & Wintle, 2008) or "ad-hoc planning" (Ruhl & Fischmann, 2011). Lance Gunderson and Stephen Light (2006) refer to this as "passive adaptive management," and point out how it is fundamentally different to the originally conceived comparative experimental design. Although adaptive management remains a popular framework among managers, it is only one of many tactics that scientists themselves are using for dealing with uncertainty, including the tactics I will identify and explore in this book.

Nevertheless, many agencies working to restore salmon to the Columbia River Basin have adopted an adaptive management approach. In his 1993 book, *Compass and Gyroscope,* Kai Lee, an academic and former board member of the NW Council, advocated for incorporating adaptive management into planning and policy in the region. Adaptive management has since become an important component of recovery planning in the basin and is a major feature of the strategy for most monitoring programs (Bouwes et al., 2016). Yet despite its conceptual value, many people also believe that adaptive management in the Columbia River Basin has not delivered on its promises to balance uncertainty while still taking action (Blumm & Paulsen, 2012; Doremus, 2001; Ruhl & Fischmann, 2011; Volkman & McConnaha, 1993). This seeming failure of adaptive management requires a deeper look at how natural resource science and policy adapts to and deals with change.

In order to deal with vexing management issues, environmental managers often call for adaptive management, yet it is still unclear exactly what this recommendation entails. This is often further confused by the lack of acknowledgement of the interaction between social and ecological systems and their *combined* capacity for adaptation and/or resilience. Management goals for achieving resilience, for instance, would be aimed at strengthening the ability of a system to recover from and resist a shift into a new state, whereas goals aimed at climate change adaptation seek to anticipate and prepare for a shift to a new state (Harris et al., 2006). This lack of clarity often permeates work in the field of environmental management. For example, in a review of 113 papers on conservation management, most of the recommendations for dealing with climate change were vague, 70 percent of them suggesting overly general principles such as "becoming more flexible." The review authors concluded that "climate change adaptation work" is still largely at the "idea" stage (Heller & Zavaleta, 2009, p. 18). Others have also pointed out that—especially when dealing with novel or no-analog ecosystems—management techniques generally provide little or no guidance (Starzomski, 2013). Although adaptive management is often referred to, explicit outcomes or actions for habitat conservation and planning are not usually included (Bernazzani, Bradley, & Opperman, 2012). In terms of "what to do" about climate change then, restorationists are mostly left to "figure it all out as they go."

Indeed, "embracing uncertainty" is still a major issue, even when using adaptive management (Keith et al., 2011). One reason for this may be because it requires policy-makers to embrace an unknown outcome—and a potential failure of the prescribed action—something that is antithetical to "good" decision-making. Another problem may be that modeling for critical uncertainties themselves is underdeveloped (Keith et al., 2011). Yet, even when these problems are taken into consideration, scientific work, which involves designing an experimental framework and monitoring change, can be difficult or impossible when environmental baselines are actively shifting and cross-scale interactions are complex. Therefore, revisiting adaptation in environmental management from a science studies perspective is warranted because it can provide insight into how scientific practice itself can be adaptive to meet the needs of changing social and environmental factors.

These frameworks for adaptive management, governance, and law all attempt to address the difficulties of dealing with uncertainty at different scales and locations. Yet, while they each highlight the critical role that knowledge—and science—plays in facilitating adaptation and resilience, none of them explicitly examine adaptation within scientific work or epistemic communities themselves. In order to advance these ideas, *Anticipating Future Environments* takes these concepts as a starting point for thinking about how socio-ecological systems can deal with uncertainty and climate change and then goes further, pushing for an "adaptive restoration" that considers the practices of scientists and managers themselves as they adapt to environmental change.

A PLACE CALLED "TRANSDISCIPLINARITY"

This book is framed using critical perspectives that may seem at odds with the applied efforts of natural resource managers and policy-makers. In these pages, pragmatic approaches to management meet the often radically deconstructing potential of science studies. The theories, perspectives, and ideas presented here are not usually brought together in polite conversation, and doing so renders theoretical and practical tensions that, in all likelihood, cannot be resolved. This is not a mistake, but it is the purpose and the point: to sit in the uncomfortable borderlands of this in-between, "transdisciplinary," or even "undisciplinary" (Haider et al., 2018), place in order to see what emerges here. While this engagement of—what may seem like—conflicting ideas is not an accident, as so much in life; this project also came about not entirely by plan. The idea came about after taking a leap into a world that has always troubled me: the world of trying to "manage" nature. My aim is not to sweep these tensions under the rug, but rather to openly engage with them.

The problematic relationship and perceived divide between controlling and using nature and preserving it has been an ongoing tension throughout my own life, and my personal history has no doubt influenced my research for this project. In large part, this story has been about facing these tensions head on. Before delving into the world of research, I spent my time engaged in radical and utopian environmental and political movements that included building intentional, sustainable, communities and spaces. Like many others, my environmentalism was something that emerged organically as an extension of my fascination with and love of nature and the outdoors. As a young person, I didn't connect it to any particular ideology; it just seemed "right." Yet my nascent, and unnamed, environmentalism was always overshadowed by a practical conundrum—my livelihood was dependent on my family's logging business. I spent my childhood in the mill-yard behind our house, playing on descriptively named logging equipment, like "chippers" and "fellers" and "skidders." After school finished for the day, I would often take jarring rides in the old log trucks with my father to the clear-cuts and mills. I knew, at an early age, where our livelihood came from, and I also saw the pride that my family took in the work they did—they saw themselves as woodsmen and hunters, making a living from the earth. While those ugly clear-cut scars and the birds made homeless (that we ended up caring for) when their forest was removed disturbed me, I learned to live in this in-between world. I didn't reconcile or surrender my values but held onto them in this intermediate place. For me, this feeling is similar to doing work that draws on different disciplines and ideas—what has come to be called inter- or transdisciplinary work: it's uncomfortable, and there is no "right" answer (although there is almost always a more "just" or "equitable" one).

Interdisciplinary work is defined as research that integrates concepts and tools from two or more disciplines (Cronin, 2008). One way that transdisciplinary work is understood is that it takes research outside of the academy to collaborate or co-produce knowledge with nonscholars (Walter et al., 2007). While this present work is undoubtedly interdisciplinary because it brings together concepts from the history of science, philosophy of science, social studies of science, and sustainability science, it can also be seen as somewhat transdisciplinary because it incorporates the perspectives of ecological restorationists through the use of ethnographic methods. Scholars in interdisciplinary fields such as cultural studies (Johnson et al., 2004) and, more recently, sustainability science have argued that "problem-based, integrative, interactive, emergent, reflexive" forms of science more accurately reflect the process through which researchers themselves deal with complexity (Robinson, 2015, p. 70). I hope that the concepts—some from vastly different worlds—that I have brought together in this work *will* be unsettling, because that is my aim. I also hope that, by bringing them together, I can introduce some ways to think differently about environmental science and environmental change that might foster creative ideas about how to confront the large-scale and complex problems that our society is faced with.

Situating the Study

This project was developed during a time when there was increasing regional and federal funding and support for climate change science and mitigation and adaptation research. During the first two years of this research, I attended conferences and workshops across the basin, observing the restoration community and discussing the many issues that restorationists were, and still are, facing when dealing with climate change in their work. The effects of climate change were being keenly felt, especially during the record-breaking, almost snow-less, winter of 2015. In many places in the Columbia River Basin, salmon returns were at an all-time low, yet, still, there was a tentative optimism. Restorationists, for the most part, believed that, with support, our society would be able to adapt to or mitigate the worst effects of climate change and that the federal government would increasingly prioritize this effort.

That bubble burst for many people when Donald Trump was elected president in 2016. The bursting of this bubble also coincided with my first year of conducting in-depth interviews with restorationists throughout the Columbia River Basin. While the topic of climate change was already politicized in much of the rural West, with President Trump entering the White House, an all-out war was seemingly declared against environmental protection and the science that supported it. The highest office in the nation was regularly denying the existence of climate change and was openly hostile toward climate change research, mitigation, and adaptation. Many programs working on topics related to climate change began having their funding cut. I met scientists who were fired or demoted in an effort to end their research programs. Even in agencies where this wasn't happening overnight, long-term cuts were being threatened from the federal level. As I worked through my interviews with restorationists, morale was low and fear and rage often ran palpably under the surface.

This was certainly an interesting time to be conducting a study on how scientists were dealing with climate change. Their main concern often centered on a lack of political and institutional support for their work at a fundamental level. Some participant scientists joined marches and protests, openly engaging in advocacy and political action to keep climate science and climate adaptation on the federal agenda. This unexpected political shift undoubtedly colored the interviews and observations in this study. Most often, this change was illustrated through expressions of anxiety, sadness, and exhaustion, as people saw their life's work being swept under the rug, ignored, or literally going to waste in a federal political atmosphere openly hostile and toxic to anything associated with climate change, and even science itself. Fortunately, many of the state, tribal, and local agencies and organizations felt this effect less acutely, as political support for environmental issues remained and remains high at regional, state, and local levels in most of the Pacific Northwest. Regardless, everyone I spoke to was aware of the political shift, and most people brought it up in interviews. This all points to a fundamental fact of using an ethnographic approach: that it is a *situated* method, and although it is possible to illustrate concepts that might be transferrable, it is first and foremost grounded in the time and place that it is conducted.

Finally, and importantly, despite the crucial perspective of the Native American peoples in the United States and Canada who are intrinsically affected by the topic of this work, I am a first-generation immigrant to the United States and the Pacific Northwest, and so will not speak for Indigenous peoples. Instead, my engagement with Indigenous voices is only through relaying perspectives of people that I spoke to in interviews and conversations. Some of these interviews were with tribal members who also work as restorationists, and who generously offered their time. It is important to note that these are individual perspectives and, as such, do not necessarily represent the opinions of other members of their tribe, or Indigenous people in general.

METHODS

My research combines the interpretive, qualitative methods of grounded theory (Charmaz, 2005), archival and policy analysis, and ethnographic methods, including interviews and participant observation. Situational analysis, which draws on grounded theory, provided tools that highlighted controversies over knowledge, which are understood to be highly contextual (Clarke, 2005). This aligns well with Donna Haraway's (1991) "situated knowledges" concept, which contextualizes knowledge historically, and is committed to recognizing the "truth" within embodied claims. This framing is important because it supports postcolonial perspectives and ensures that actors are not silenced—a particularly important consideration in the study area of the Columbia River Basin where traditional and Indigenous knowledge plays a critical role in decision making and restoration in the region and where colonialism has been—and still is—a powerful force driving access to resources and the ways that we know and frame those resources.

In the case of this research, the "situation" that was analyzed encompasses both the historical development of the field of salmon habitat restoration and restoration ecology, as well as the current practices of fisheries biologists and restoration ecologists, habitat restoration managers, and people and institutions involved with developing salmon habitat policy. In order to capture emerging practices and strategies for dealing with climate change, I paid special attention to those people who are on the front lines—actually conducting restoration research and monitoring activities, and therefore making epistemic choices. In a region as large as the Columbia River Basin, a study like this one necessarily encompasses a large number of people. Therefore, a large-scale ethnographic endeavor like this one aims to sample a cross-section of actors from different sectors and locations in order to develop a broad picture, while also going in-depth in certain locations to gain a more detailed perspective of the issues. Themes as they emerged were integrated into future in-depth interviews, in order to "test" them out with participants and triangulate—that is, validate or dispute—them.

To gain a depth of understanding beyond the individual perspectives captured in the interviews, the study also included observations from major regional conferences and associated conference workshops on ecological restoration and salmon recovery themes. At these conferences, I was able to observe emerging topics of concern and engage in informal discussions with people about how they were struggling with and through these issues in their field. I also took part in virtual workshops and webinars held regularly among practitioners in the region. I was generously invited to join restorationists in the field and I was shown around many of their restoration sites; and I participated in management site visits where experiments were being conducted or management decisions were being made. These visits were invaluable for contextualizing the issues that participants discussed in interviews and allowed me to further triangulate emerging topics and concepts. Finally, historical documents at the University of Washington Special Collections, as well as policy documents from organizations in the basin such as the NW Council and the Bonneville Power Administration (BPA) were also analyzed. From this work emerged patterns and themes that identified both drivers of change and responses to change in restoration practice. In order to analyze these patterns and themes, I turned to the theories and concepts on science and change from science and technology studies (STS) as well as the history and philosophy of science.

UNDERSTANDING ADAPTATION IN SCIENCE

By examining knowledge production and scientific work itself, we can approach adaptation in governance and resource management from a different angle. As outlined above, much of the literature on scientific and policy change in environmental management is framed in terms of adaptive management. The literature on adaptive management is based on the contention that as science is conducted and management conditions change, new information will become available and actions may need to be adjusted. Yet, despite its seemingly simple formula, in practice, adaptive management has been difficult to fully implement (Blumm & Paulsen, 2012;

Craig & Ruhl, 2014; Doremus, 2001; Ruhl & Fischmann, 2011; Volkman & McConnaha, 1993). By understanding how science is or is not adaptive, we will be better equipped to consider what it would mean to facilitate or support adaptation in science, so that those working at the forefront of adaptation will be more prepared to confront environmental change.

Change in the Columbia River Basin is not a new phenomenon. Environmental and social changes have been drastic and ongoing since European settlement, as well as the millennia before, as Indigenous peoples also shaped and reshaped the landscape. Science in the basin, too, has always been changing, with scientists constantly adapting their practices to different drivers of change. This book brings attention to the ways that scientists themselves conceptualize these changes—both the drivers of change, and their responses to these drivers. For this, I turn to theoretical concepts from the field of STS, which includes historical, philosophical, and social perspectives on science. These concepts help tease apart the dynamics of stability and flexibility in scientific work, including changes in scientific practices and epistemic cultures and virtues.

Theories of change in the field of science studies provide some ways to conceptualize these shifts in epistemic communities. In *The Structure of Scientific Revolutions*, Thomas Kuhn (1962) famously shifted the study of science to consider what scientists actually "do" when they are conducting scientific work. Instead of the more common, teleological argument that conceptualizes science and technology as constantly progressing as discoveries are made and knowledge is gained, he argued that there are periods of "normal science" which are upset by scientific revolutions, or "paradigm shifts." According to Kuhn (1962), when scientists in a given field are conducting "normal" science, they share a "paradigm," including the same beliefs, understandings, theoretical frameworks, models, and methods. Scientists go about solving "puzzles" using these shared worldviews and ways of working. Eventually, however, unsolved puzzles accumulate into "problems," creating a period of crisis, which necessitates changes to the paradigm. Through concerted effort, this crisis is eventually resolved, and a new paradigm is constructed that enables normal science to continue. Kuhn's (1962) work was important because he not only offered an early framework for understanding change and stability in science, but also a way to explain how different paradigms can coexist, even when they are at odds with one another, or in his words, "incommensurable."

Science studies scholar, Ian Hacking (1992) also looked at stabilities in science, and found that disciplines develop bodies of theory, scientific apparatuses, and methods of analysis that are constantly being rewoven and adjusted to maintain continuity. He uses the analogy of a rope to describe science: there are multiple strands and, while any one strand can break, the rope may still hold (Hacking, 1992). These strands can be theoretical, experimental, or instrumental technologies. So, while some parts of an epistemic endeavor may change, others may stay the same. Instead of transforming though a Kuhnian paradigm shift, science may remain stable even if a new theory or technology becomes available. And, instead of entering a major shift, scientists may slowly adapt to technological innovation, data needs, or exogenous changes that require new theories or focus.

For ecological restorationists, climate change is one of these gradual shifts or adjustments, and it is currently underway. From a science-studies perspective, Silvio Funtowicz and Jerome Ravetz (1993) argue that, because society now requires science to constantly describe risks and confront environmental crisis, it operates in a "post-normal" state. In a "post-normal" mode, scientists recognize the complexity of nature and move beyond a positivist and modernist vision of technoscientific progress (Funtowicz & Ravetz, 1993). In their conceptual model, unpredictability, contradiction, and complexity are assumed, and this "enriched awareness of the functions and methods in science" is welcomed (p. 740). Similar to adaptive management, in a post-normal science, values are explicit, and uncertainty is therefore both recognized and "managed," instead of ignored. Funtowicz and Ravetz (1993) argue that a post-normal science has the potential to transform science itself into a more democratic force: by recognizing multiple ways of knowing that go beyond scientific expertise, and thereby addressing risk and the environment in a more just way.

As I described above, while work on adaptive environmental science and management often discusses the need for knowledge to support adaptation, the details of this dynamic are unclear. Fortunately, I can borrow concepts from STS that allow me to draw out and bring shape to some of these complexities in scientific work. For example, in returning to the questions: what kinds of knowledge would be adaptive, and what would the knowledge infrastructures, institutions, and organizations that facilitate this adaptation look like? What kinds of practices would signify that this adaptation is occurring? The following concepts will help give nuance to the explorations of these questions in the chapters to come.

Practices

In one sense, ecological restorationists are adapting to basic, ontological changes as their object of research—the river environment—is altered by climate change; but the adaptations that follow include changes in their practices. The practices of scientists themselves are constantly remaking the social fabric, or culture, of science. Therefore, science can also be understood as both performative and temporally emergent through practice (Pickering, 1995). Examining scientific practice highlights the role of individuals and their actions in either remaking or extending—in other words, *changing*—a particular scientific culture. Pickering (1995) describes this process of cultural extension in science in what he terms the "mangle of practice," whereby the goals of scientific work are constantly "accommodated," revised, or "tuned" as scientists encounter "resistances" through practice. Resistances could be things that don't quite fit into a model, technologies that fail, or difficulties encountered when capturing data. According to Pickering (1995), the dialectic process of "resistance" and "accommodation" is what drives practices to change, and thereby culture also to shift. As scientists cope with the uncertainty of a shifting climate, they also shift their epistemological practices: they find ways to deal with "resistances" to their efforts by adjusting to or "accommodating" them. Therefore, practice is one place where we can observe the adaptation of scientific work, and these adaptive practices will be highlighted throughout each of the following chapters.

Cultures

Epistemic cultures are "cultures that create and warrant knowledge" (Knorr-Cetina, 1999, p. 1). Epistemic cultures include the ways and norms of working within and achieving expert status in a particular scientific field, and, like all cultures, they change. For example, in Karin Knorr-Cetina's

(1999) ethnographic work on epistemic cultures, she goes beyond describing and defining a discipline and instead looks at the "epistemic machinery," or organization of expert systems that are a precursor of scientific work. To do this, she examines the different empirical approaches, collections of instruments, and social mechanisms that make up an epistemic culture. She demonstrates that disciplines are not homogenous and describes how they often contain overlapping but distinct epistemic cultures (Knorr-Cetina, 1999).

Ethnographic methods like the ones used in this study are particularly useful for examining both practices and cultures, and researchers in STS have developed a significant literature devoted to studies of them. These studies use ethnographic methods like participatory observation and interviews to examine the performative work that scientists do within laboratories (e.g., Knorr-Cetina, 1999; Latour & Woolgar, 1979; Law, 1987) or in field sites (Kohler, 2002). Following in this vein, I focus on the way that culture is employed in scientific practices, allowing me to observe strategies that restorationists are using to adapt to environmental change. While Knorr-Cetina was concerned with the epistemic cultures found within large-scale experiments and laboratories, other researchers have focused on field science (Geissler & Kelly, 2016; Kohler, 2002; Latour, 1999).

The "large-scale experiment" (Gross, 2010) of restoration in the Columbia River Basin has similarly developed its own cultures of field science, as different restoration enterprises have emerged in the Columbia River Basin over the past several decades—including the engineering-based and process-based restoration cultures that will be examined more closely in later chapters. As I will explain in chapters 1 and 2, one reason for this is that scientists are responding to certain regulatory metrics that have exerted political force as they have become institutionalized. The epistemic cultures and expert systems that result from this interplay of law and science are therefore unique and situated, and they are also adapting. Although the different cultures of restoration—engineering- and process-based—may have often been at odds, in chapter 4, I will explore points of convergence and shift between the two cultures, as they come together to address climate change through the emergence of different restoration tactics.

Virtues

Looking more closely at the motivations of the people who are actively transforming epistemic cultures and practices, we can see that they reside in specific "forms of the scientific self" (Daston & Galison, 2007, p. 4). Lorraine Daston and Peter Galison's (2007) history of how the concept of "objectivity" in science has transformed through time demonstrates how scientists hold to particular virtues, or "ethos" that are both "ways of being" and "ways of knowing." Like epistemic cultures, these virtues are historically situated and change through time. Epistemic virtues and norms like truth, objectivity, replicability, or creativity become important tools for scientists to employ, and to employ at distinct historical moments. While these virtues are only one aspect of a much broader culture, they play an important role in epistemic work, shaping the field and the way that science is conducted at particular points in time.

It is important to point out that epistemic virtues differ from a concept like simply being a "virtuous" scientist, in the sense of having competence or expertise. Instead, they refer to "normative codes of conduct ... bound up with a way of being in the world" (Daston & Galison, 2007, p. 40). For instance, Daston and Galison (2007) find that epistemic virtues such as "truth-to-nature," "mechanical objectivity," and "trained judgment" evolve over time. While these virtues can deem a scientist "virtuous" during a particular point in history, it is the ways in which the virtues change, playing distinct and different roles in knowledge-making during different historical periods, that makes them interesting to a study such as this. During different periods of history, scientists have performed different virtues at different times, when possessing them was viewed as an important part of the scientific endeavor, thereby allowing us to observe another facet of adaption in science.

In this study, changing norms related to what is deemed "good" and "bad" science can be found in the ways in which participants discussed their own—and other people's—restoration work and scientific practice. Chapter 6 homes in on these changing virtues in the restoration community of the Columbia River Basin, where practitioners are increasingly coming to value experimentation and flexibility in the face of uncertainty. These are especially underpinned by an epistemic virtue of efficiency, or what Daston and Galison (2007) refer to as "pragmatic efficacy," which enhances restorationists' ability to act quickly in the face of a changing environment. This adaptation extends even further to value anticipation itself as a virtue. This is something which has been found in other epistemic communities that shift to look to the future: a moral injunction to anticipate the future, while conducting ongoing scientific work, emerges as an epistemic virtue (Adams, Murphy, & Clarke, 2009).

Yet, it is important to keep in mind that this is not a seamless process; and as these new virtues emerge, they also align with particular scientific practices—which can often be at odds with one another. For instance, the epistemic virtues of precision and replicability are often contradictions (Daston & Galison, 2007). Similarly, a need to experiment with and adapt restoration practices in the Columbia River Basin is in tension with the certainty of the "precautionary principle" demanded by the ESA—where species must be recovered at "whatever the cost." As I will describe, disciplinary principles and virtues within engineering, for instance, can also conflict with restorationists seeking to exploit ecological processes.

Knowledge Infrastructures

Studies of knowledge infrastructures in STS examine the structures that support scientific work. While knowledge infrastructures include institutions, organizations, collaborations, and norms, they also encompass material structures like laboratories, research centers, and computer networks. Knowledge infrastructures are therefore included here as an object of study because they capture not only the more abstract things such as metrics, standards, and networks of people, but also the scientific instruments, models, and data networks and repositories that are concrete, yet often invisible and taken for granted (Bowker & Star, 2000). Knowledge infrastructures support the work that scientists do and influence the way that science is applied (Bowker & Star, 2000). Infrastructures are especially important to consider in this study of adaptation in scientific work because they enable scientific work to occur. As such, they can be designed and maintained in ways that facilitate or hinder adaptation (Ribes & Polk, 2015). Because their materiality and rigidity have legacy effects, their design has consequences for the science that results (Edwards, 2010). Yet, in the context of environmental management, these infrastructures must be able to adapt to changing objects of research that come with the shifting socio-ecological conditions and trajectories of climate change. The knowledge infrastructures that are in place today will shape future scientific capabilities as well as the programs and resources dedicated to salmon recovery in the Columbia River Basin. In order for these programs to remain successful, it is therefore important to understand how knowledge infrastructures can be adaptive in a time of unprecedented environmental change.

Salmon habitat restoration in the Columbia River Basin is supported by a large-scale knowledge infrastructure. This knowledge infrastructure includes the individuals, their collaborative practices and norms such as standards and routines, as well as the physical spaces and materials such as cyberinfrastructures and field stations that support their work. This infrastructure is considered "large" because of its geographic and temporal scale, ambitious goals, large economic investment, and coordination among many actors (Edwards, 2003). In chapter 5, I will show how ecological restorationists are adapting, or acclimating, knowledge infrastructures by changing the ways they collect data and choose standards in order to keep their work relevant in a changed and changing climate.

MOVING FORWARD

These concepts from the field of science studies outline some of the sites where we can observe change in knowledge production. By looking at the scientific practices of restorationists, the knowledge infrastructures that support those practices, and the institutions and organizations that facilitate or hinder adaptation, we can better understand what an adaptation to environmental change in science actually looks like.

I begin in chapter 1 by historically situating the scientific work of habitat restoration in Columbia River Basin. By following the story of Harlan Holmes, a fisheries scientist whose work began in the 1920s, I describe how the basin has been developed and the ways in which the science that has evolved there has been co-produced to meet the needs of a particular imaginary of the river, what historian Richard White (1999) calls an "organic machine." This demonstrates one of the many ways that science is not merely an abstraction but also transforms the material world to meet particular imaginaries.

Chapter 2 follows the evolution of ecological restoration more broadly, and then gives an overview of particular perspectives on restoration in the Columbia River Basin. In doing so, I also introduce many of the conflicts and tensions that have emerged, including the differences between engineering-based approaches to restoration and ecological processes-based approaches. The institutional and organizational histories in chapters 1 and 2 set out as the basis for understanding restoration in the Columbia River Basin; but in chapter 3, I begin to delve into what restorationists actually do in practice. Here, I follow the work of a large-scale restoration project and the people who developed and implemented it. Through their work, I describe some of the issues they encounter when faced with climate change, including the problem of "shifting baselines" (Pauly, 1995).

The following three chapters each develop different concepts for understanding adaptation in scientific work. Chapter 4 uses the example of beaver restoration to describe the concept of "emergence," whereby engineering-based and process-based restoration cultures are coming together to foster experimentation and the emergence of new ideas. Chapter 5 outlines the concept of "acclimation" whereby restorationists are slowly adjusting their monitoring activities and attendant knowledge infrastructures so that their work can remain useful in a climate-changed future. Chapter 6 develops the final concept, "anticipation," by following the work of ecological and environmental modelers and the ways in which modeling can be used to help tame epistemic indeterminacy and envision potential futures. While each of these concepts hails from a different disciplinary perspective, they share two qualities that make them useful in this study: they can all be observed through ethnographic methods and they are all sites where change is negotiated in scientific practice. Therefore, by paying careful attention to how the epistemic community of ecological restoration relates to these concepts, strategies for dealing with change also become apparent.

These three adaptations that I outline are by no means the only ways that restorationists are coping with change, but they were important themes in this research and they serve as a conceptual framework for beginning to consider adaptive scientific practices, knowledge infrastructures, and institutions and organizations. They are kinds of "adaptive epistemologies," a term that was first used by Mieke van Hemert (2013) to describe "an approach to biodiversity restoration . . . which implies tolerating a certain degree of epistemic indeterminacy" (p. 75). I would like to use this definition as a starting point and build on it to not only include the "epistemic indeterminacy" that van Hemert identifies, but the myriad other ways that scientists adapt their work in the face of change. Some of these tactics are the concepts I introduce in this book. I am calling these *adaptive epistemologies:* the ways that scientists adjust their scientific practices, the knowledge infrastructures that facilitate these practices, and the institutions and organizations that are co-produced by them.

I hope that these concepts can serve as reminders that scientists are constantly adapting to social and environmental change. I expect that they may also provide some conceptual space for people engaged in environmental management and scientific work to reflect on how climate change is influencing their own practices and how it might do so in the future. This is a step away from the promises of modernity—technoscientific solutions and certainty—that instead looks to the actions of individuals and organizations that are working on habitat restoration *today* in order to capture the changing and inherently social nature of epistemic work.

While conducting the fieldwork for this study, ecologists, managers, and policy-makers all expressed a need to understand the changes occurring throughout the basin, and they wanted to know how climate change might likely affect their science and their practice. But what I heard most often was a need for practical advice and tools for incorporating climate change into their work. I cannot offer many management suggestions or ways to make the river or the fish more physically resilient. But after speaking to people and observing their work across the basin, I *can* tell the story of restoration science in the basin—what has happened in the past, how it has changed, and what restorationists are doing now as they adapt to climate change in their work. While outlining the specifics of how this might be done in practice is not my main objective, the ideas laid out here were developed with users in mind. I want to offer scientists, managers, and policy-makers a conceptual space to think about where they might focus their own efforts in strategizing for climate change in the context of scientific work and environmental management. I hope that the ideas offered here will be of value to the people who are dealing with the immediate effects of climate change in their own work.

The effects of climate change in the Columbia River Basin and the Pacific Northwest are apparent and they are growing. From drought and wildfire to extreme flooding, participants in this project, as well as the wider community of people concerned with restoring salmon to the rivers and streams of the region, have been trying to figure out what to do. Yet while *un*certainty about what to do seems to be growing exponentially, *certainty* that environmental conditions are changing is also growing. Those working on salmon habitat restoration can see the ways that environmental conditions are shifting—and it is affecting their work. This puts restoration practitioners in a pivotal place for instituting adaptive changes in the short term. They are sensitized to these environmental changes and they are taking action: shifting their restoration practices by facilitating the emergence of alternatives, acclimating their methods to allow them to continue to move forward in their work and, at the same time, anticipating the future.

1

A Science for the Columbia River Basin

IN THE SUMMER of 1923, Harlan Holmes embarked on his second field season as a biologist for the US Department of Commerce Bureau of Fisheries. His work would eventually take him throughout the Columbia River Basin, traveling by train, bus, ferry, canoe, and whatever rides he could hitch along the way from cannery owners and fellow scientists. In later years he brought his own car for this journey, but this led to its own problems: spending many of his field days as a shade-tree mechanic, trying to keep the engine running as he drove around remote areas of Washington, Oregon, and Idaho. Holmes's task was to survey the Columbia River Basin: to find places where salmon still thrived, where they were no longer present, and to locate sites on which to build hatcheries and field stations for scientific work. He was a fisheries scientist, and as a trained biologist he wanted to measure fish, mark fish, and count fish in order to understand what their life cycles entailed and how changes in the environment affected the different species.

He has been described by those who knew him as both "inquisitive" and "completely dedicated to his work" as a scientist (Atkinson, 1988). It is also clear from his field journals that he was a conservationist, and the declining salmon populations worried him. Holmes saw this negative trend as an issue to be remedied, and he brought science as his tool in this early recovery effort. His career spanned the 1920s into the early 1960s, where he worked for both the Bureau of Fisheries and the US Fish and Wildlife Service, and later the US Army Corps of Engineers. This was a time when the "salmon crisis" of precipitously falling stocks was just beginning to be recognized. To figure out the best way to remedy the observed population declines, those in charge called upon science. In this way, Holmes was a forerunner to the ecological restorationists working on the revitalization of critical habitat and salmon conservation today. His daily field notes—stuffed with jottings, plans, photos, and fish-scale samples—describe his scientific practices as well as his conservation ethic as he went about his work.[1]

In the early twentieth century the Columbia River was dominated by large canneries, and their fish wheels unceasingly scooped unlimited quantities of fish from the river (Taylor, 1999). In his first two field seasons, Holmes traveled from cannery to cannery throughout the lower and mid-Columbia, trying to enroll workers into his "marking" project. His goal was to establish a monitoring program so that he could follow the migration patterns of the "marked" fish. This would allow him to understand some very basic information about salmon behavior: he simply wanted to know where they were coming from and where they were going to. At this time, little was known to science about salmon life cycles or genetics, and there was very little scientific understanding about distinct populations, runs, or "races" (Taylor, 1999). Holmes also wanted to know exactly how many fish actually survived the canneries' fish wheels on their way up the Columbia. Holmes was one of the first fisheries biologists tasked with figuring all of this out.

The first step in his plan was to find cannery workers and fishers who would join his scientific enterprise to mark and measure fish. At this time there was very little documentation about the fish that were being caught by commercial fishery operations (Taylor, 1999). On his way to the field, he stopped and bought twenty-five yardsticks to put in the canneries as he visited each one: "Such places are particularly in need of measuring sticks," he wrote (H. Holmes, 1922). In addition to setting up measuring stations, he developed his own traps to catch fish at particular points along the river, scooping them up and putting them into ponds or tanks that he created on site. There was almost no infrastructure to support Holmes's work, and he assembled all of his field labs from scratch. Going about his work in this way, he was setting up a "home" for science that was outside of the lab and in the field (Geissler & Kelly, 2016). This meant that many of his days were spent moving equipment or constructing traps and ponds to hold fish. Holmes also spent a lot of time negotiating with hatchery and cannery owners in order to secure places to establish his longer-term experiments. Throughout all of this practical, physical work, he was building an infrastructure to support his science in a landscape that lacked any.

Having identified some locations to conduct his experiments, Holmes set to work scooping up fish. But this didn't always go according to plan. In fact, he didn't have much of a plan to go by, as he was conducting novel experiments that hadn't been done before. He didn't have any expert, manual, or textbook to consult. Instead, he tried different methods, adapting them as he went. Holmes often lacked equipment, and this too, he improvised along the way. For example, on May 22, 1923, Holmes spent the day near Lake Quinault, north of the Columbia River Basin, catching fish to mark for his experiments. He recorded in his journal:

> Made a circular net about 4 ft in diameter and took it down to the trap at the outlet of the lake where we easily caught several hundred fine looking fish about 3½ inches long. Brought them back to the hatchery in a milk can, pouring water continuously on the way. (H. Holmes, 1923)

But the next day he found his previous day's work lost:

> MAY 23: Got up around 6 this morning to catch more fish. Went to the hatchery and found about half of those placed there last night, dead. I caught a few more and transported them carefully from the boat house to the hatchery, a distance of a few hundred feet. We spent most of the morning operating a short seine and caught several thousand fish, the most of which were placed in live tanks but a few were taken to the hatchery. I marked a few

about noon. It was noticed that many scales were removed in handling and a couple of them died soon. Losses in the hatchery were large even with the most carefully handled fish. (H. Holmes, 1923)

His Lake Quinault field station, near the Washington coast, seemed plagued with problems. Eventually he determined that many of the fish in his containment net had been infected with a parasite that rendered his experiment useless. Soon after this incident, he abandoned that project for the season.

Further inland, on the main stem of the Columbia, Holmes continued to face issues with keeping fish alive in adverse conditions. The lack of infrastructure in place to support his field stations was taking a toll on his efforts. As he traveled to visit his hastily constructed sites, he would often be forced to leave others in charge. Again, returning to his field station in Quinault, he found many of his marked fish dead or dying. The manager he had left to care for the fish didn't seem to realize that the food he was feeding them was completely rotten:

> Mr. Larsen seems to see nothing wrong with food which to me seems badly spoiled. This liver, which left Portland Monday and was brought to the hatchery Thursday P.M., was placed in cold running water and ground when needed. The last was ground on Friday morning and kept in a bucket immersed in running cold water. By Sunday the ground liver had a distinct odor and today it not only had a strong odor but showed evidence of forming gas. Had new food not arrived this old food would have been used tomorrow. (H. Holmes, 1923)

As these journal entries demonstrate, lack of trained personnel with the expertise needed to support his scientific work was endemic. Holmes struggled to get all of his workforce to use rigorous methods and take reliable measurements. "I am impressed by the unreliability of certain figures," he wrote in reference to one employee's reporting, "Such approximations as this would be worse than worthless even if listed as such" (H. Holmes, 1923). While these difficulties beleaguered his field stations, he was still forced to forge quickly ahead for lack of time: the marking methods he was developing required him to complete his work before incoming fish runs crossed over with the outward-migrating, ocean-going fish. Despite numerous setbacks and failed experiments, Holmes worked to refine his methods and by the end of his field season in 1923, he and his team had marked 101,000 fish, recording: "the extra 1000 to cover losses resulting from overcrowding and lack of sufficient flowing water on May 18, and what loss there has been immediately following marking. This likely more than doubly covers the loss prior to liberating" (H. Holmes, 1923). While Holmes's early work for the Bureau of Fisheries may seem crude and lacking rigor, he was building upon previous biological knowledge and practicing his work according to the norms of a field biologist of his discipline in his time.

Holmes's scientific work in the basin tells us a lot about how science has changed, it also coincides with and parallels developments that were happening throughout the basin. A lot has changed in the Columbia River Basin both before and since Holmes's first scientific expeditions in the 1920s. In the early 1900s, at the time Holmes was penning his journals, the river was in the process of being developed on an extensive and rapid scale. While a comprehensive history of the river's development and the colonial and ecological legacy of this development is more than this book can offer, in this chapter, I will provide a brief background of the development of the river, the basin, and the conflicts that have resulted.[2] This background is important, not only for understanding the setting for this tale of a single scientist's work; it also provides a necessary foundation for comprehending the complex conflicts at the nexus of science, policy, and environmental justice that exist in the basin today. These interactions demonstrate what science studies scholar, Sheila Jasanoff (2004) has termed the co-production of science and law—whereby science is used to support legal action, and the resulting policies, in turn, create a pivotal role for science. Organizations can co-produce science in straightforward ways, such as allocating funding through policies and institutions. Yet, throughout this book, I will argue that the institutions and organizations in the Columbia River Basin co-produce science in more subtle ways as well, as they require science to meet particular sociotechnical and environmental imaginaries for the river.

THE LEGACY OF DEVELOPMENT

The Columbia River Basin encompasses a diverse landscape roughly the size of France. The headwaters of the Columbia River begin in the Rockies and the Cascade Mountains, draining dozens of sub-basins and major tributaries including the Snake, Salmon, Clark Fork, Willamette, and Yakima Rivers in the US, as well as the Kootenay and Okanagan Rivers in British Columbia. Almost all of the precipitation that falls on Idaho, Oregon, and Washington east of the Cascades, as well as large areas of Montana and British Columbia, will drain through this system. This water eventually reaches the Pacific after funneling through Portland and a 47-mile-long estuary at the river's mouth.

Indigenous people have fished the Columbia River Basin for salmon for at least 10,000 years (since the last ice age) (Butler et al., 2019) or maybe more appropriately, what the treaties and the tribes call, since "time immemorial." Since this time, the salmonid populations of the Columbia River Basin have gone from abundance, with tribal annual catches in the millions of tons to extinction—or near extinction—in many parts of the basin (Goble, 1999).

To the Indigenous people in the Columbia River Basin, the destruction of the salmon populations is a profound cultural, spiritual, economic, and subsistence tragedy that is the ongoing legacy of colonialism. According to Nez Perce tribal elders: "one of the greatest tragedies of this century is the loss of traditional fishing sites and Chinook salmon runs on the Columbia River and its tributaries" (Landeen & Pinkham, 1999). There are many different tribes and cultures in the Columbia Basin, and their beliefs are diverse. Yet many of them hold creation stories that feature the salmon, considered as a gift from the Creator as a First Food (CRITFC, 2014). This profound loss has been actively countered since the 1850s by a series of treaties to reserve hunting and fishing rights, as well as lengthy and ongoing battles both in rivers and in courts to preserve and strengthen these rights (Cosens, 2012).

While many people today consider the large hydropower dams on the Columbia and its tributaries to be the main cause of harm to anadromous salmon populations, many different flavors of overexploitation and development have led to the low numbers of salmon that we see today. To many

Euro-Americans who were arriving in the region during the mid-1800s, the salmon of the Columbia River represented a kind of "gold-rush," but it wasn't until the technology of fish canning was introduced the 1880s that the "fish-rush" really began (Goble, 1999). The canneries could export as much fish as they could catch, leading to fisheries technologies that would scoop as many fish as possible from the river using nets, wheels, and traps. Production reached a maximum in 1911, with 50 million pounds packed, but this unsustainable fishery eventually collapsed, and by the 1970s, the fish runs were virtually annihilated (Goble, 1999).

In addition to overfishing, other unsustainable land-uses also impacted fish runs, including pollution from mining and dredging, overgrazing, and logging, as well as agricultural development and large-scale irrigation projects (Goble, 1999; White, 1995; Taylor, 1999). At the same time as hydropower development was taking place in the basin, it was also being transformed into what was imagined to be an "Irrigated Eden" (Fiege, 1999b). Large-scale irrigation systems often drove or went hand in hand with large-scale dam projects, as extensive areas of the Columbia and Snake River Basins were converted from desert-shrub lands into irrigated farms. This began as early as the 1860s, but by the 1920s, the Bureau of Reclamation (BoR) had built hundreds of miles of canals and irrigated hundreds of thousands of acres of arid land (Fiege, 1999a). This network drastically changed the hydrology of the rivers, by shifting water to different areas: creating new wetlands where none existed and dewatering others. These physical changes were in addition to the straightening of complex streams, a rise in water temperatures, introduction of pollutants, and trapping of fish in irrigation canals (Fiege, 1999b). While dams are often central to environmental debates about salmon survival, these multiple impacts of development must all be taken into account when considering salmon habitat restoration today.

THE DAM DEVELOPMENT ERA

Despite these other changes to the river, dam development was probably the most devastating change to the habitat of anadromous fish species. The Columbia River and its tributaries were transformed during the first half of the twentieth century, as large-scale hydropower and irrigation projects sought to put the power of the river to work (White, 1995). Environmental historian Richard White (1995) describes how the Columbia River was turned into an "organic machine," a kind of hybrid "energy system" which still "maintains its natural, its 'unmade' qualities" (p. ix). White's history is provocative in that he uses the concept of energy to highlight the ways in which the histories of both humans and nature are entwined:

> The flow of the river is energy, so is the electricity that comes from the dams that block that flow. Human labor is energy; so are the calories stored as fat by salmon for their journey upstream. Seen one way, energy is an abstraction; seen another it is as concrete as salmon, human bodies, and the Grand Coulee Dam. (White, 1995, p. ix)

I quote White at length here because he is pointing to a dynamic that runs throughout this book: while science can often be seen as an abstraction—an idea or mode of working that is somewhat removed from the material—like energy, science also involves human labor and resources, and it transforms the material world. Scientific labor makes and remakes the physical river through the type of information that is gathered and the way it is presented to policy-makers, who then employ that science as a basis for their actions. In turn, policies support particular forms of science through funding or regulations that require specific forms of knowledge or intervention to interpret environmental change. In other words, like dams, science is social just as much as it is technical—both science and dams are sociotechnical and they are co-produced.

As a result of the engineering accomplishment of dam-building and irrigation on a massive scale, the Pacific Northwest went through a period of rapid economic development, becoming an agricultural and manufacturing leader—and now a hub for tech—that is still reliant on Columbia River energy and irrigation. Today, hydroelectric dams on the Columbia River and its tributaries provide over half of the electricity-generating capacity for the Pacific Northwest region of the United States, and this proportion is due to grow with the closure of thermal (coal) plants and increases in efficiency in energy consumption (NW Council, 2013).

The first phases of hydropower development in the river basin began as early as 1888, but it wasn't until the 1930s that the large, major dams began to be constructed (Goble, 1999). There are several social movements that convened to muster the economic and political will to develop these large dams. In addition to the technological utopianism of the 1930s, the project was heavily influenced by an era of progressive "high-modernism" (Scott, 1998). This effort reflected a trend in technological optimism occurring across the country as the government sought to put nature to work through large-scale investment in infrastructure and regional planning (White, 1995; Hirt & Sowards, 2012). Mainly constructed during the dam-building era beginning with the New Deal in the 1930s and on into the 1970s, there are now almost 130 dams that are used for hydroelectric power, 31 of these are large-scale dams built across the main stem of the Columbia and its major tributaries such as the Snake (Goble, 1999). While the economy of the Pacific Northwest may have benefited from this inexpensive, renewable, and mostly federally owned energy source, these dams, along with development and exploitation of the rich natural resources of the Pacific Northwest, have irrevocably altered the ecosystems and devastated the anadromous fish populations that call them home.

One of the first large dams in the Columbia River Basin that was completely impassible to anadromous fish was the Warm Springs Dam, built on the Malheur River in Oregon in 1919. By 1920 the first comprehensive plans for building dams throughout the region were being drawn up. Between the 1930s and the 1970s, over two dozen large dams were constructed on the main stem of the Columbia and its tributaries. These dams were built with a combination of benefits in mind, including hydropower, irrigation, and, after the 1948 flood of Portland, flood control. The year 1938 saw the Bonneville Dam completed. In 1941, the Grand Coulee Dam was finished, and it is still one of the largest dams in the world, blocking access for migrating salmon to over 1100 miles of river habitat (Goble, 1999). In 1957, the Dalles Dam flooded Celilo Falls, one of the most important tribal cultural and fishing sites in the region (Barber, 2005). Dams continued to be built through the 1970s, with major dams constructed on almost all of the main tributaries. This included some of the later dams to be built: the Hells Canyon Dams, which blocked salmon habitat in vast areas of the Upper Snake River and its tributaries. In fulfillment of the Columbia River Treaty, signed in 1961, several large Canadian dams were also constructed for both flood control and hydropower. In Idaho, Dworshak Dam, one of the highest in the world, was one of the last dams to be built, in 1973. With the catastrophic failure of Teton Dam in 1976, the dam-building era of the twentieth century came to an end; but only after transforming the Columbia River into one of the most regulated rivers in the world—and reducing salmon spawning habitat from approximately 163,000 to 73,000 square miles (Goble, 1999).

It is important to point out, however, that this new river, or "organic machine" was not created blindly—the consequences to salmon were known at the time (White, 1995; Goble, 1999). The Fish and Wildlife Service, for example, recommended a moratorium on dam construction because of the devastating effects that were already being noticed (Goble, 1999). This was in 1946 (Goble, 1999). Yet, as history shows, techno-optimism won out; hatcheries and fish passage, instead of a free-flowing river, were seen as the answer to the declining fish runs (Taylor, 1999; Goble, 1999).

ENVIRONMENTAL HARM AND LEGAL RESPONSES

Some may view these river-transforming projects as embodying the dream of American democracy, or even democratic socialism, of the 1930s: others, a technological hubris that unleashed devastating environmental harm and immeasurable injustices to the tribes that had relied on salmon as a First Food and who hold undeniable treaty rights to this resource. Regardless of the reasoning for this enterprise by the US government (and a few private companies), as a direct result of habitat loss from hydropower development, thirteen salmonid species are currently listed under the Endangered Species Act (ESA) as either endangered or threatened. Salmonids—both salmon and trout—have complex life cycles that can entail adult migration hundreds of miles upstream from the ocean to spawn in tributaries. Juvenile fish return to the ocean after spending the beginning of their life in fresh water. They then migrate throughout the North Pacific for two to six years maturing to adulthood before eventually returning to the same river in which they hatched. While dams were originally thought to impact salmon mainly by creating barriers to upstream migration, there is now a better understanding of their multiple effects. Dams decrease juvenile survival as salmon migrate downstream to the ocean, by increasing water temperatures and migration times, inflicting fluctuations in oxygen levels, and causing mortality through contact with dam infrastructure (Dauble et al., 2003).

The scale of the Columbia River Basin and the complex life cycle of the salmonids that live large parts of their lives in it ensure that the effort to restore endangered fish populations is highly intricate. The Columbia River is the fourth-largest river by discharge in North America; it crosses seven state boundaries and dozens of sovereign tribal nations' reservations as well as their usual and accustomed territories; and constitutes a major international transboundary river. Straddling the border of the United States and Canada, the Columbia River Basin is not only ecologically complex, but also the site of international diplomacy through the Columbia River Treaty, which includes negotiation over water rights for hydropower generation and flood regulation to protect major cities in the US as well as communities along the river in British Columbia. While the Columbia River Treaty between the United States and Canada does not mandate consideration or coordination of issues related to ecological integrity, this topic continues to emerge in transboundary talks (Cosens & Williams, 2012). Adding ecological function as a "third prong" to the treaty, which is limited to issues related to flood control and power generation, was even recommended by the US Entity in the treaty review process (US Entity Treaty Review, 2013). Any salmon restoration effort within the Columbia River Basin must therefore be coordinated across state, federal, and tribal boundaries, as well as between public and private landowners, making management and monitoring along the river continuum a highly complex problem.

Until recently, there has been little watershed or basin-wide coordination or monitoring of restoration efforts (Katz et al., 2007). Although it is difficult to know exactly how much is spent on habitat restoration across the region, it is estimated that at least $300 million USD per year is spent in the Columbia River Basin (Bernhardt et al., 2005; Katz et al., 2007; Rieman et al., 2015). In 2016, the Northwest Power and Conservation

Council alone allocated $274.2 million to recovery, with $117.9 million of that going directly to habitat restoration and protection (NW Council, 2017). Many still wonder if this is enough, if it is too late, or if restoring habitat is actually making a difference for populations that are still declining (Katz et al., 2007).

Most salmon species also migrate throughout the northern Pacific Ocean. This compounds the spatial and jurisdictional complexity of the river continuum with that of the ocean and has made any attempt to restore salmon throughout their entire habitat and life history a truly complicated political and practical problem. Regulation, recovery, and restoration of salmon species must be organized between international marine waters. In the United States, the National Oceanic and Atmospheric Administration Marine Fisheries Service (NOAA Fisheries), along with federal and state fish and wildlife agencies and coordinated intertribal bodies such as the Columbia River Inter-Tribal Fish Commission (CRITFC) are major players tasked with implementing the restoration effort. Meanwhile, organizations such as the Northwest Power and Conservation Council help coordinate and distribute habitat restoration funding for the hydropower mitigation efforts of the Bonneville Power Administration. In addition to these larger organizations, individual consulting and engineering firms, tribal, state, and local, natural resource departments, and community groups also play important roles in the recovery effort.

THE ENDANGERED SPECIES ACT

For those tasked with restoring salmon habitat in the Columbia River Basin, the scope and scale of this "wicked" problem can be overwhelming. Yet, despite these difficulties, the restoration of salmon habitat in the Columbia River Basin is still proceeding. Restoration of the Columbia River Basin is fairly unique in terms of the scope and scale because it is driven by mandates from the ESA as well as tribal treaty rights. As such, it is relatively well-funded compared to restoration efforts in other river systems. Columbia River tribes were some of the first to advocate for endangered species listings: after years with only one or two adult salmon returning to spawn to their tributary headwaters, the Shoshone-Bannock Tribes put forward a listing proposal for the Snake River sockeye in 1990.

According to legal scholar Dale Goble (1999), at its most basic, the ESA "embodies a simple proposition: all species have a right to continued existence" (p. 250). This means that any modification to critical habitat of species listed by the ESA is either prohibited, or the damage inflicted must be mitigated in some form. All dams in what is now known as the Federal Columbia River Power System (FCRPS) are managed by one of two federal agencies: the Bureau of Reclamation (BoR) or the Army Corps of Engineers. A third federal agency, the Bureau of Reclamation, markets the power produced by the FCRPS. Dam management therefore constitutes a federal action subject to requirements of federal law, including—among others—the ESA. Section 7 of the ESA provides that all federal agencies must ensure that their actions not jeopardize the continued existence of a species listed as threatened or endangered, nor result in the destruction or adverse modification of its "critical habitat." These "action agencies" satisfy the Section 7 requirements through a process known as "consultation" with the appropriate federal fish and wildlife agency, either the US Fish and Wildlife Service for terrestrial species or NOAA Fisheries for marine species and anadromous fish (including Columbia River Basin salmon and steelhead).

Where it is determined that a federal action is likely to adversely affect a listed species or is likely to result in the destruction or adverse modification of critical habitat, the appropriate wildlife agency prepares a "biological opinion" (BiOp) which determines whether the action will result in jeopardy or adverse effects. If the agency determines that jeopardy would result, either the action cannot proceed without violating Section 7 or the action agency must implement "reasonable and prudent alternatives" (RPAs) to the action in order to avoid jeopardy—in other words, mitigation for the damage that the dams inflict on salmon must take place.

After NOAA Fisheries (then known as the National Marine Fisheries Service) listed the Snake River sockeye salmon as endangered in 1991 (the first salmonid to be listed in the Columbia River Basin) and the Snake River fall and spring/summer Chinook as threatened in 1992, the BoR, the Army Corps of Engineers, and BPA initiated consultation with NOAA Fisheries to determine the effects of dam operations on these species. In the following years, NOAA Fisheries listed ten additional Columbia River Basin anadromous salmonids as threatened or endangered, which required additional consultations to determine the effects of dam operations on those listed species. Including the initial consultation in 1992, NOAA Fisheries has completed eight BiOps on the effects of dam operations on Columbia River Basin salmonids. Various interest groups have challenged all eight, alleging an array of legal problems with the opinions. Of course, this legal framework for protecting species from extinction has not been this simple to carry out in practice: all of the BiOps since 2000 have been found to violate the ESA for various reasons.

ADAPTIVE MANAGEMENT AS AN ATTEMPT TO DEAL WITH CONFLICT

In order to deal with uncertain environmental and social conditions while at the same time addressing the need to protect endangered species, many government agencies have embraced adaptive management. As discussed in the introductory chapter, adaptive management is an attempt to deal with scientific uncertainty by continually revising management actions in light of new knowledge and experience (Holling, 1978; Walters & Riddell, 1986). In 1993, Kai Lee, an academic and former board member of the NW Council, explored the interface of science and policy in the Columbia River in his book *Compass and Gyroscope*. He described a need for both adaptive management and political deliberation. Managers and scientists who were eager to explore the possibilities and nuances of incorporating adaptive management into policy-making welcomed this contribution (Lee, 1993).

The Bush administration tried and failed to have the dams themselves excluded from the analysis of impacts to listed salmon in the 2004 BiOp. Although adaptive management has been both explicitly and implicitly used within the BiOps and the operation plans for the river, it did not become an explicit legal issue until the 2008 BiOp. The revised 2008 BiOp attempted to remedy many of the problems with previous BiOps by providing funding for habitat restoration and mitigation options, such as hatchery and transportation, that were reasonably certain to occur at least in the short term. After reviewing the 2008 BiOp, the Obama administration created an adaptive management implementation plan (AMIP) designed to implement the reasonable and prudent alternatives (RPAs) more effectively in an adaptive management framework. While a legal challenge to the 2008 BiOp was pending before Judge Redden, the agency requested a stay of the proceeding and a remand to allow it to incorporate the AMIP into the BiOp. The agency then issued a supplemental BiOp with the AMIP in December 2010.

Unlike previous BiOps, this 2008/2010 BiOp included specific mitigation projects and funding for the first five years of its ten-year lifespan (i.e., 2008–2013). After 2013, the agencies would rely on the monitoring and studies in the AMIP to determine what mitigation measures should be implemented during the second five-year period. Because the action agencies would develop them based on the experiences of the first five years of its lifespan, the BiOp could not identify what the future actions or projects would be. Nor, given the inherent uncertainty of restoration and mitigation efforts, could it ensure that *effective* projects could even be created. Consequently, although Judge Redden agreed that the 2008/2010 BiOp satisfied the ESA's requirements for the 2008–2013 period, the lack of mitigation that was "reasonably certain to occur" for the 2013–2018 period rendered that aspect of the BiOp arbitrary and capricious. For that second five-year period, the adaptive management measures were simply a "promise to figure it all out in the future" (NWF v. NMFS, 2011). At least as applied in that BiOp, Judge Redden identified a danger in substituting adaptive management, and the uncertainty inherent in that approach, for substantive decision-making and planning, and concrete, specific mitigation projects.

In May 2016, after Judge Redden retired, a new judge remanded the 2014 Supplemental BiOp that NOAA Fisheries created in response to Judge Redden's remand of the 2008/2010 BiOp. The 2014 BiOp failed in part due to its failure to "properly analyze the effects of climate change," and because it was "inconsistent" in its "treatment of uncertainty" (NWF v. NMFS, 2016). In his order, Judge Simon recognized this as "picking and choosing" between which uncertainties to emphasize, something that has discredited science in other contexts (Herrick & Sarewitz, 2000). While adaptive management was once seen as a way to deal with uncertainty, the level of uncertainty within these BiOps proved too much for the court, even when it was part of a formal adaptive management plan (the AMIP).

As this episode demonstrates, despite its ambitious goals and theoretical value, adaptive management has often failed in practice. Since Lee's efforts to bring adaptive management to management of the Columbia River, other natural resource scholars have interrogated adaptive management and found that time and again, adaptive management fails to deliver on its promises to balance uncertainty and decision-making in natural resource management (Blumm & Paulsen, 2012; Volkman & McConnaha, 1993; Ruhl & Fischmann, 2011; Doremus, 2001). The problems with adaptive management, uncertainty, and risk become especially clear in the story of the intense litigation over the BiOps for endangered and threatened salmon recovery (Morse, 2012; Doremus, 2001; Blumm & Paulsen, 2012; Blumm, Thorson, & Smith, 2006; McLain & Lee, 1996). Critics of NOAA Fisheries' efforts to implement adaptive management have dubbed it a "watered-down" version of the principle that is more like "ad hoc contingency planning" than "learning by doing" (Ruhl & Fischmann, 2011, p. 426). Other observers were even more critical, calling the agency's efforts outright "deception," claiming the agency misused scientific authority in the form of adaptive management to convince the public that recovery and restoration are taking place, when in fact they are not (Blumm, Thorson, & Smith, 2006). While adaptive management is one attempt to address uncertainty by acknowledging and planning for the potential failure and subsequent improvement and reimagining of mitigation efforts, it has thus far been rejected by the court.

A SCIENCE FOR DAMS

These fundamental issues with adaptive management are important to consider when thinking about how scientists may adapt to deal with climate change. In order to do this, let us take a step back from this legal (and scientific) story. While the sole purpose of the ESA is to prevent extinction at "whatever the cost," it is also important to recognize that the loss of fisheries has fundamentally impacted the treaty rights of Native American tribes in the region. As described earlier, because of hydropower development, Columbia Basin tribes have suffered the loss of a critical First Food along with fishing sites of irreplaceable cultural and spiritual significance (Barber, 2005; Pearson, 2012). A growing demand for and recognition of tribal treaty rights (Cosens & Williams, 2012) is adding to the call for mitigation and habitat restoration from both a legal and an environmental justice standpoint.

The transformation of the river and salmon habitat during the twentieth century required a profound simplification of natural complexity (Hirt & Sowards, 2012). Nature was altered on a massive scale. The scientific and natural resource institutions and organizations that were set in place to support this transformation still influence the material possibilities in the basin today, as hatchery science and fish passage technologies became the preferred ways to tackle salmon decline (Taylor, 1999). As we have seen, habitat loss due to hydroelectric infrastructure, as well as agriculture, forestry, and municipal development, have all been major factors in declining populations of salmonids in the basin; yet habitat protection was not considered a priority until the late twentieth century. Despite this, it is now widely recognized that habitat restoration *is* critical to promoting salmon survival, particularly in the rearing and spawning phases of their life cycle (Stanford, Fissell, & Coutant, 2006).

Yet, during the first part of the twentieth century, scientific efforts were not focused on preserving fish habitat; instead, the interest was in development, and in finding technological fixes such as getting fish around dams by using fish ladders or establishing runs of hatchery fish (Taylor, 1999). The establishment of this scientific focus can be seen by returning to Holmes's survey work, funded by the 1927 Rivers and Harbors Act, in the headwaters of the Columbia. The act requested that the Army Corps of Engineers develop a "comprehensive plan" for the Columbia River Basin. The comprehensive plan, titled "Columbia River and Minor Tributaries," included a survey of habitat; but its main purpose was to locate areas where hydropower, irrigation, navigation, and flood control projects should be developed. Also known as the "308 Reports," the plan eventually resulted in the transformation of the Columbia River into a series of reservoirs, destroying or blocking much of the habitat that Holmes was busy documenting in the 1920s.

As Holmes's career progressed, he spent less and less time up in the headwaters in summer talking to locals, surveying the areas where fish were spawning and counting "bluebacks," or sockeye. Instead, he moved on to work as a biologist for the US Fish and Wildlife Service and carried out most of his studies at the first dam fish meet when traveling up the Columbia: Bonneville. He spent the remainder of his career trying to determine the effects of the dam on juvenile and adult salmon and experimenting with ways to diminish these impacts. He developed fish passage technologies for many of the large dams, as well as numerous screening devices to protect juvenile fish. By the end of his career, he was working almost solely on Bonneville Dam passage experiments. Holmes's scientific work had come to be meeting the needs of river development, and technological fixes had become the answer to the dams' impacts on salmon. While he started his career concerned with habitat depletion, his scientific work shifted toward fish passage technology along with the demands of river development and the information required to address the issues framed by society.

As Holmes traveled further into the headwaters in the years after his initial fieldwork for the Bureau of Fisheries in 1923, he began talking to people about constructing fish passage around dams. The BoR was just beginning work on the Yakima Project, a massive engineering and irrigation effort that would convert the Yakima Valley into one of the most productive fruit-growing areas in the United States. To do this, in 1917, the BoR dammed the Keechelus River. On his field trips right after the dam was built, Holmes met with people who were eager to find ways to keep fish coming up the river and spawning in the valley. They also wanted to find ways to develop technology to stop fish from entering and getting trapped by irrigation canals. He met with residents and fishers near Cle Elum, Washington, and discussed fish passage around the dams. People there were concerned that the development was harming one of their cherished resources: the fish runs.

In 1924, Holmes's field season took him further afield. He drove his car up into the Okanagan River Valley of Washington and British Columbia. There, he collected scales from salmon to take back to his lab to determine the differences between fish populations, the main purpose of this research trip. But, in addition to biological descriptions and fish scale samples, his field diaries are also filled with resident's fears about the changes in abundance of fish runs as well as the shifts they were beginning to notice in the timing of the runs. In many of the locations he traveled to, the effects of river development were being both seen and felt, and residents were clearly noticing changes in the river environment itself. Holmes recorded finding many streams dewatered. For example, while traveling through the Okanagan Valley, he took a side trip to look at Shingle Creek. "It is about like the other creeks in this valley," he wrote, "not much more than a brook, the greater part taken out for irrigation. There were no fish in sight at the places where I observed it, but I did not go to the mouth" (H. Holmes, 1924). Holmes met with similar sights in the Methow Valley of Washington, where he discovered streams that were once rich salmon spawning grounds diverted for irrigation and agricultural development.

In 1927, Holmes traveled throughout the Salmon and Snake River Basins of Idaho, encountering many of the same issues. He "learned at Idaho City in the evening that many Chinooks spawned in Boise River before dams were constructed" (H. Holmes, 1927). But locals were particularly angry about Sunbeam Dam. This dam was built in 1910 by a private mining company in order to provide electricity for its operations. Its construction effectively blocked salmon from returning to a large part of the Salmon River Sub-Basin (USFS, 2018). Holmes's field notes about how locals felt about the dam are particularly telling:

> It is interesting to note that everyone in this part of the country is very much opposed to the dam and is very free about stating that they would like to see it blown out. It is a wonder someone has not done so long before now. (Holmes, August 12, 1927)

The locals did get their way in 1934, when the dam was, in fact, "blown out," at the request of the United States Forest Service (USFS) after the mining company failed to produce a profit and abandoned their operations.

The conversations recorded by Holmes demonstrate that at least some non-tribal communities in the early part of the twentieth century clearly cared about having salmon in the rivers. The quotes in Holmes's diaries also show that they could see how the development of the river was having a detrimental effect on salmon abundance. But Holmes and his colleagues went one step further and saw the potential for restoration of salmon habitat to areas with recently blocked fish access or dewatered streams. In the upper reaches of the Salmon River, Holmes was particularly interested in documenting where fish were spawning, and he recognized the same type of high-quality habitat that has now been prioritized for restoration almost a century later. He wrote in his 1927 field journal:

> Took my time going down Bear Valley. Looked along the River (a very small stream at the head of the valley) at several places but saw no evidence of salmon or spawning grounds. Took a picture looking toward the head of the valley. The valley is a level and comparatively flat meadow several miles wide in some places. The river windes [sic] its way in a very crooked course through this marshy flat. It must be a wonderful nursery grounds for aquatic life upon which young salmon might feed. (Holmes, August 12, 1927)

It would take almost a century before valley floors like this one would be recognized for their critical role in salmon recovery and would be restored to the highly complex, braided streams that form in wide valley bottoms—what are now called "Stage 0" floodplains.

Yet Holmes's work on documenting habitat did not continue; instead, his science was needed elsewhere. He became central to one of the early legal controversies surrounding dam operation. He was especially interested in understanding how the pressure created by dams impacted juvenile fish as they passed through the turbines, and he created novel experiments to enable him to test the effects, both in the lab and in the field, marking and collecting fish above and below the dams. In 1952, he wrote a report titled "Loss of Salmon Fingerlings in Passing Bonneville Dam as Determined by Marking Experiments." Up until this point, it was generally assumed that juvenile fish passed through the dam turbines or over the top of the dams with few problems. This report, and especially the statistical methods that Holmes used, became critical in establishing whether or not the dams were going to have to admit responsibility for damaging juvenile fish. The paper went through many rounds of reviews, and the Army Corps of Engineers (Corps) were particularly unimpressed with the conclusion that the dams contributed to large loses of juvenile fish. The Corps attacked Holmes's statistical methods, particularly the chi-square tests, and drafts of the document were sent to respected statisticians across the country to determine whether his conclusions were warranted. Lt. Col. L. W. Correll of the Corps was adamant that they were not disputing the damage to fish by the dams but were instead disputing the statistics in the report. He wrote in a letter:

> Again, I wish to state, however, that the Corps of Engineers would not object to an adverse report on the effects of the Bonneville Dam on the salmon fishery so long as the adverse findings are based on factual and conclusive data. I believe that the answer to this question is so important and far reaching to the future of all the resources concerned that no answer should be declared or implied until the supporting facts are conclusive. (Lt. Col. Correll, October 19, 1952)

As we will see throughout the story of science in the Columbia River Basin, the Colonel was right: the controversy over which "supporting facts" are, in fact, "conclusive" is still the topic of debate, and the answer is still "important and far reaching to the future of all the resources concerned." Although this controversy over fish ladders is just one of many in which science was enrolled, science would continue to play a key role in determining the impacts of dams, as well as the solutions to these impacts, whether these solutions were hatchery propagation or habitat restoration.

A RIVER CO-PRODUCED

Legislation such as the ESA provides an example of what has been termed the co-production of science and law. Co-production describes a situated and iterative process, where the "ways in which we know and represent the world (both nature and society) are inseparable from the ways in which we choose to live in it" (Jasanoff, 2004). While there are undoubtedly people that tend toward either *producing* science or *implementing* restoration actions, in this study many restorationists stand with a foot firmly in both worlds, and through their practices, and the epistemic cultures, virtues, and institutions that support them, they all take part in co-producing restoration in the Columbia River Basin. Co-production in environmental management is an important dynamic to consider because the way in which it is done has material implications for how the environment is managed. This is because *knowledge about nature* is co-produced along with societal actions *to manage* the environment and order nature.

Co-production provides a useful idiom for conceptualizing the relationship between science and institutions, especially legal institutions such as the ESA (Jasanoff & Wynne, 1998). Science can also reflect political goals through different civic epistemologies and institutions that facilitate the production of knowledge in order to legitimate the state (Jasanoff, 2005). As the story of Holmes's research in the Columbia River Basin demonstrates, while scientists produce knowledge in terms of epistemic norms and values, the science they produce is nonetheless often created in order to meet the needs of particular societal norms and values that are expressed through legal frameworks. Scientists and institutions can co-produce science in straightforward ways, such as allocating funding through policies and institutions. For instance, the BPA delivers most of the funding for habitat restoration in the Columbia River Basin as mitigation for the effects of the dams, and in many ways, BPA funding defines what kind of knowledge gets produced. In coordination with the NW Council, BPA not only funds monitoring programs, but also prioritizes projects, metrics, and methods of restoration. Yet, organizations like the BPA also mean that the research funding structure of restoration ecology in the Columbia River Basin is unique. For example, several scientists told me that it is generally an "unwritten rule" that grants from other science funding agencies such as the National Science Foundation are almost impossible to get for projects in the region because it is assumed that the BPA will be funding local fisheries-related research. Funding is also spatially determined by the fact that the BPA is only responsible for mitigating the effects of dams; so areas below the Bonneville Dam, such as the estuary, were not recognized as being affected until very recently— and even now only to the tidal high-water mark. In these ways and many others, institutions and organizations are a determining factor for what kinds of scientific work get done in the basin.

Yet, as Harlan Holmes's evolution as a scientist demonstrates, institutions and organizations co-produce science in more subtle ways as well.

While he began his career in the 1920s, documenting salmon decline and exploring their headwater habitats, Holmes's science was eventually called upon to seek solutions to fish passage around dams instead. This represents a shift in the way the river was conceptualized: from a free-flowing river to a highly regulated one. But it is also a profound shift in the *type* of knowledge that was needed to understand the river and the salmon in it.

The legal regimes within the basin set out the need for particular kinds of knowledge. Since the time of the first ESA listings, the epistemic community of ecological restoration in the basin has grown rapidly to meet the mandate to recover salmon. While the ecological complexities and political hurdles to restoration can seem overwhelming, people in the conservation and restoration community have nevertheless been galvanized by the opportunity that listing afforded. Rosemary Furfey, a recovery coordinator at NOAA Fisheries, has been witness to the rise of the restoration effort in the Columbia River Basin. She reflected on the uniqueness of this situation:

> We actually have the opportunity in the Pacific Northwest. We still have wild fish. A lot of people don't have that, like in New England where the Atlantic salmon is very endangered, and almost gone in Ireland and Scotland. We actually have an opportunity and I think it is very inspiring.

To a restoration ecologist, recovering salmon to the Columbia River Basin through habitat restoration is a once in a lifetime chance to put their scientific practice to work.

The development of the Columbia River Basin has clearly been controversial. Yet, as I will show in the following chapter, ecological restoration itself has also had a contested history. Opinions on all sides of the scientific and managerial debates contribute to the creation of a diverse epistemic community that has varied through time and across contexts. While the way in which the epistemic community has developed is reflective of the unique mandates for salmon recovery under the ESA, these kinds of conflicts exist wherever environmental regulation calls on science (Bocking, 2004; Alagona, 2013). The metrics used to meet these mandates, and the epistemic communities that they rely on, co-produce both restoration and the science that supports it. This is true not only of the science conducted in the Columbia River Basin but of any environmental knowledge produced in order to understand and manage the environment the world over. And, it is the environmental and sociotechnical imaginaries of societies that drive this co-production, stifling or promoting the potential to have a river managed with a more environmentally just future imaginary.

2

River Restoration in the Columbia River Basin

THE RIVER RESTORATION Northwest Symposium has been an annual occurrence for the past twenty years. Many of the more than four hundred people at the conference have been attending for over a decade. It is an event where established colleagues come together to discuss their projects, students attend to learn new skills and network for jobs, engineering firms arrive to showcase their latest technologies, and the "restoration industry" comes together to share new ideas. This year was no different, and the "new idea" on many people's minds was something called "Stage 0."

For decades, river restoration has hinged on the idea that an alluvial—or sediment-driven stream—develops as a single channel, with water reaching flood stage and topping banks every few years. This single-channel framework, or "Channel Evolution Model," is still the most common way to classify and understand both past and future changes in stream morphology (Schumm, Harvey, & Watson, 1984). Many restorationists use these stream models when they plan restoration projects, allowing them to describe a specific stream's structure and function, how it evolved, and how it may change given different conditions. The Channel Evolution Model follows the advancement of a stream from Stages 1 to 7 as it develops due to hydrogeomorphic processes such as incision or downcutting, and aggradation or sediment deposition. Another, somewhat controversial classification system, "Natural Channel Design"—or the "Rosgen" system (Rosgen, 1994)—has also dominated restoration design and similarly follows a single-channel framework. Dave Rosgen's system has been particularly influential, and its uptake has been linked to certification programs that have facilitated privatization of the restoration industry in some regions, blurring the line between commercial and academic research in problematic ways (Lave, 2012; Lave, Doyle, & Robertson, 2010). The controversy over whether or not the Rosgen system is the best way to conceptualize change in stream channels is still a heated one in the restoration community. But regardless of the model used, until recently all of them have assumed single-channel rivers and streams to be the norm.

These single-channel models are now being questioned (Walter & Merritts, 2008). According to a new framework, the "Stream Evolution Model," the prevalence of single-channel streams is a human-created condition resulting from land use: it is partly an artifact of development that drained wetlands, ditched fields, and constructed homes on historic floodplains (Walter & Merritts, 2008; Cluer & Thorne, 2014). As discussed in chapter 1, this type of development was a common occurrence in the Columbia River Basin. The Stream Evolution Model adds a "Stage 0" and a "Stage 8" to the evolutionary cycle of a stream. These stages represent multi-threaded or "anastomosed" streams in floodplains that are typically inundated by floodwater several times a year (Cluer & Thorne, 2014). Streams like these are now thought to more closely resemble predevelopment conditions, including areas influenced by beaver activity (Bouwes et al., 2016). New field studies are now considering floodplain streams more closely, and as the results of some of these studies are now becoming finalized, researchers are finding that Stage 0 streams provide more ecological benefits than single-channel streams. These benefits include increased habitat diversity and physical resilience, as well as amelioration of the high temperatures that can be detrimental to cold-water species like salmonids (Cluer & Thorne, 2014).

Stage 0 was a big deal at the annual symposium. The new concept was forcing many restorationists to rethink restoration goals, the engineering methods used to reach those goals, and the metrics used for measuring success. To be sure, not everyone was on board with this new concept. Some people were skeptical of its utility—or its necessity—and were debating it both openly on stage and in the sidelines and hallways of the conference. But for many at the conference, the words Stage 0 were providing a new way to think about their own restoration work and the role that they played in altering stream physiology. As one of the speakers at the symposium stated: "It's a paradigm shift."

During one session at the conference, talks given by some of the pioneers of the Stage 0 concept were followed up by presentations on preliminary results from projects that had recently implemented the concept in their designs and monitored the effects. Practitioners were excited to see the results and assess for themselves whether Stage 0 would help them in their own projects. To many, it made sense that restoring streams back to this earlier stage would garner more ecological response, especially in terms of the goals that were emerging as important for mitigating the effects of climate change: increasing groundwater interaction and creating a diversity, or mosaic, of habitat types all within close proximity one to the other (Beechie et al., 2013). As I will describe in chapter 4, restoration using beavers and beaver dam analogs (BDAs) is fast becoming an important tactic, and Stage 0 also represents a "re-beavered" landscape. Stage 0 therefore aligns well with larger conceptual shifts taking place in the field of restoration. But, most importantly, restorationists at the conference were excited because Stage 0 gave a name to something that they had been working around for a while: restoring floodplain connectivity using natural processes and moving away from the more heavily engineered, channel-centric approaches that have been the more commonly used restoration tactics in the past.

Restoring a river in order to recover a species, whether salmon in the Columbia River Basin, or any other species in diverse ecologies around the world, requires drawing from expertise across many fields: from engineering to biology to ecology to geomorphology. River restoration is about more than just "fixing" a broken stream, it also involves everything that connects to that stream and the organisms that rely on it—in this case, the endangered salmonids as they move throughout their complex life cycles. When people in the field refer to the work of "restoration" they are usually casting a broad net. They may be including riparian and streamside habitat: the wetlands and forests and estuaries and reservoirs that salmon pass through at different times in their (non-ocean) lives, as well as the stream morphology: the arrangement of rocks and gravel and debris that forces the stream move in a particular way. Restoration, therefore, also encompasses the geology of the river itself, along with the flow of water: the element that is most often in greatest need of being restored. As one restorationist at the conference said, their job is to "re-complexify a simplified river." So, what exactly goes into this "re-complexifying," and, more critically for this study—which focuses on scientists and their practices—how do scientists understand and go about this activity?

What follows in this chapter is a brief history of the field of restoration ecology and ecological restoration, including its relationship to natural resource management. I then delve into some of the issues that specifically apply to stream restoration, including the major themes and debates that developed in early ecological science and how those themes have influenced the ongoing development of the field. I then shift to explore how the field of restoration has evolved specifically in the Columbia River Basin. One driver is the debate between engineering-based and process-based restoration. In the following chapters, especially chapter 4, I will explore how this debate—or what could also be characterized as a scientific controversy—has proven productive for the emergence of new and adaptive ways of working within a climate-changed environment.

ROOTS OF ECOLOGICAL RESTORATION

The tale of Aldo Leopold (1949) and his experiments in prairie restoration at the University of Wisconsin are an often-repeated origin story for ecological restoration. But there are other possible precursors for the restoration enterprise. One of the most obvious, but neglected, origins of restoration are the Indigenous ecologists and hunter-gatherers, who practiced land management through a deep understanding of ecological processes—constituting something which could be called restoration in today's scientific jargon (Allison, 2012). An alternative origin story finds the roots of ecological restoration in early forest management in Europe, dating to Germany in the 1300s or England in the 1600s (Glacken, 1990

[1967]); Allison, 2012). River restoration itself has its own origin stories, some of which draw from landscape architecture and horticultural design (Egan, 1990). These multiple histories point to the various ways in which humans have for millennia worked to repair damaged ecosystems, either for the purpose of managing the landscape for increased productivity or for aesthetic or cultural reasons. Regardless of where we begin our history, restoration has certainly existed for a long time in one form or another.

In the American West, restoration as a scientific endeavor began in earnest in the early 1900s, as researchers and managers at land-grant institutions attempted to restore forests and rangelands to productive conditions after logging, grazing, or devastating wildfires. This was partially done through the work and visions of George Perkins Marsh and John Wesley Powell, who viewed the West as suffering from overexploitation of both grazing lands and forests (Hall, 2005). Some of this work was carried out through the US Forest Service Great Basin Experiment Station, established in 1912 (Hall, 2005). This is a different origin story for North American restoration than the more common one that attributes it to Aldo Leopold.

Restorationist Stuart Allison (2012) believes that the origin story of Leopold is so strong today because it links restoration with a "land ethic" of restoring nature for its own sake, rather than for exploitation like that of the early US Forest Service. The Leopold story contrasts with the goals of timber harvest or rangeland use, as in the case of the earlier restoration work by Marsh and Powell. Thinking about these different perspectives on the origins of restoration in the American West is interesting in relation to ecological restoration for salmon recovery efforts in the Columbia River Basin because here the restoration is taking place, first and foremost, to prevent the extinction of salmon. Yet at the same time, there is the underlying assumption that *both* harvest *and* preservation are important reasons to restore populations. This serves as an example of where dichotomous views of nature—nature for society versus nature for nature's sake—begin to break down. When teasing apart the purpose and goals of salmon recovery and salmon habitat restoration, we find that the goals are often intertwined. These debates about the sources, and even the *soul*, of ecological science, can be found humming along in the ecological restoration taking place in the Columbia River Basin today.

Regardless of its exact roots, the field of ecological restoration that we now find in the Columbia River Basin partially grew out of the ecological awareness that was developing in the mid-twentieth century. The expansion of the Ecological Society of America in the 1970s and 1980s helped to launch both the discipline of conservation biology and the discipline of ecological restoration. Ecological restoration has since developed into its own field, with journals, societies, and conferences: all of the markers of a fully fledged epistemic community (Haas, 1990). The Society for Ecological Restoration defines ecological restoration as "an intentional activity that initiates or accelerates the recovery of an ecosystem with respect to its health, integrity and sustainability" (Society for Ecological Restoration, 2004). While this definition seems straightforward, it is anything but. Each of the words within it have been debated at length, and some restorationists still disagree with parts of the definition. Further, "restoration ecology" is often differentiated as the *science* of ecological restoration (Jordan, Gilpin, & Aber, 1987), while the broader field of "ecological restoration" includes not only scientists, but also managers, laypeople, and practitioners. Some people even see restoration ecology as the more "scientific" or "academic" endeavor of the two. Debates differentiating between these restoration endeavors have certainly been carried out in the two main journals of the field: *Restoration Ecology* and *Ecological Restoration*. Yet in reality, many practitioners and scientists span both worlds and a clear division between the science and its application is difficult to maintain. Further, as restoration science has increasingly engaged the public in restoration projects, the field has developed into a more transdisciplinary endeavor, effectively transforming itself to encompass an even broader community of practitioners (Gross, 2010). Following this, and the broad reach of restoration in the Columbia River Basin, I use the term "restorationists" to refer to anyone involved in the greater restoration endeavor, including scientists, managers, engineers, and practitioners of various sorts.

Yet, beyond the American natural resource paradigm, the broader history of ecological restoration as a field of study has had its own controversies. Restorationists, ecologists, and philosophers vigorously debate the purpose, goals, values, and meaning of restoration and its normative relationship to management, science, and the public (Bradshaw, 1987; Higgs, 2003; Jordan, Gilpin, & Aber, 1987). Historians of ecology such as

Sharon Kingsland (2005) and Donald Worster (1987) have explored how conflicting ideas about the best way to manage the natural world are evident from within the roots of ecological science itself. These histories demonstrate how particular ideologies and values form the basis for ongoing conflicts and divergences within the field of ecological restoration today.

For example, Worster (1987) identifies two threads that run throughout the development of ecology as a scientific discipline. He named them the "Arcadian" and the "imperial" ideals. The Arcadian ideal views nature as holistic—a whole, a community, or a system. It is romantic in the sense that it views cultivating a relationship with nature as an ideal, and opposite to that lies the alienation and detachment of the industrialized world (Worster, 1987). In contrast, the imperial ideal focuses on reason. Followers of this ideal try to classify nature so that it can be understood and controlled (Worster, 1987). According to Worster (1987), both of these ideals are rooted in enlightenment philosophies that were aimed at improving the natural world in order to create a paradise on earth through the scientific endeavor. As such, they consistently reemerge in political and philosophical thought through time. Even today, these ideals influence the natural sciences and resource management in different ways. For example, early approaches to natural resource management in the US drew on both of these ideals but launched scientific and managerial projects to control national landscapes through increasing legibility—mapping and quantifying resources. Both ideals can be seen in the colonial effort of explorer-scientists who worked throughout the nineteenth century to map territory so that it could be efficiently claimed and developed by settlers and the state (Cronon, 1983; Scott, 1999; Worster, 1987). The ideals can also be found, as described in the previous chapter, in the technocratic optimism that drove the development of the Columbia River Basin hydropower system and the pursuit of an "irrigated Eden" (Fiege, 1999b). From this perspective, the development of the Columbia River Basin and our scientific understandings of it are embedded within these shifting scientific and natural resource paradigms, which have changed through time and hail from different, often conflicting, roots and ideals.

The relationship between ecological science and natural resource management has been a contested one, and untangling these conflicts is a starting point for understanding how change in these scientific fields has occurred and might adapt in the future. Throughout this book, I will tease out these different paradigms by highlighting different kinds of restoration, and demonstrating how environmental science and ecological restoration in the basin have come to focus on a particular environmental imaginary (Peet & Watts, 2002) of the river. In a similar way to ideals, imaginaries are collective, socially constructed ideas about the world, and can drive the belief in, creation of, and unfolding of particular futures. There are particularly poignant imaginaries that influence nations (Anderson 2006), environments (Peet and Watts, 2002), or sociotechnical assemblages (Jasanoff, 2004). Similar to using the idiom of co-production, to directly consider imaginaries in science and technology builds on the previous chapter, and demonstrates how science can enable particular ways of understanding and regulating the environment. Ecological restoration, in other words, has not developed as an isolated scientific activity, but it has evolved to meet the needs of the natural resource management imaginary of the American West—and in turn, it has shaped it (Cronon, 1992).

DEVELOPMENT OF HABITAT RESTORATION IN THE COLUMBIA RIVER BASIN

Back at the River Restoration Northwest Symposium, some people felt that classifying streams in this new, Stage 0, way puts restorationists not only in a better position to be able to understand how streams function, but also to be able to restore them more effectively and efficiently by letting natural process do much of the work. In one conference talk, for example, a restorationist pointed out that the earlier, channel-centric approaches might have remained dominant for so long because channels were easier to "count." The length of a single stream channel can be measured using a tape or other means. Multiple intersecting and diverging streams that fan out over a floodplain make this task much more difficult. Further, the success of many restoration projects is measured in relation to recovery goals which are set to meet the metrics of the Endangered Species Act and its critical habitat requirements, and often involve counting numbers of pools, numbers of structures, or miles of stream restored.

The complex, braided streams in a highly dynamic alluvial floodplain make quantifying the success of Stage 0 restoration projects extremely difficult. But by putting a "name" to and measuring the results of this new restoration goal—Stage 0—people are able to employ it more easily in their projects, their designs, and their monitoring.

So, what does it mean to "count" a river, or to make a restoration project "count"? And, who is demanding this counting? As I explained in the previous chapter, regardless of how scientists would like to measure and monitor their projects, the metrics for success have not been determined solely by science, but they have often been co-produced with legal frameworks. As science studies scholar Sheila Jasanoff points out, "the realities of human experience emerge as the joint achievements of scientific, technical, and social enterprise: science and society, in a word, are co-produced, each underwriting the other's existence" (Jasanoff, 2004, p. 17). Although there have been significant structural and technological changes throughout the river basin, the way it is managed, the technologies for understanding it, and the scientific and natural resource paradigms and institutions that were put in place in the past still influence the river restoration being conducted today. As habitat loss and the destruction of interconnected riparian landscapes and habitats came to be recognized as a driving factor in decreasing salmon runs, scientific priorities in the basin began to shift (Stanford et al., 2006). Yet because the infrastructures for hatchery science were already firmly in place, the scientific endeavors that supported hatchery science continue to dominate in the basin (Taylor, 1999). Further, the large scale of the Columbia River Basin, the multiple jurisdictional boundaries within it, and the spatial complexity of salmon life cycles has meant that the restoration community has been forced to innovate in order to cope with these broad scales and their differing priorities.

Coincident with the early ecological restoration efforts in the US West, the 1920s to 1930s saw an increasing public awareness of the value of wetlands. Sportsmen's organizations were interested in preserving disappearing wetlands in the Midwest so that they could serve as habitat for wildfowl, and they helped push for federal regulations to protect what were previously considered "waste" lands and swamps (Fiege, 1999b). As a part of this conservation movement, the Migratory Bird Habitat Stamp Act was passed in 1934, and it still contributes to both wetlands preservation and the promotion of recreation on wetlands in the US. During the same era, riparian restoration also began to be recognized as management tool, as trout anglers witnessed declines in fish numbers (White, 1995). Yet, as described earlier, simultaneous to this growing awareness of the importance of freshwater habitats, the dam-building era in the Columbia River Basin was taking off.

The term "conservation" was originally used to encompass environmental management aimed at increasing productivity. Engineers such as those tasked with irrigating the West sought to improve nature through the creation of a pastoral promised land (Fiege, 1999b; Wilkinson, 1992; Worster, 1987). Although ecological science transitioned between 1920 and 1945, moving from a utilitarian to a preservationist perspective (Worster, 1987), the tension between utility and preservation was not resolved. Even though a more preservationist value of nature for nature's sake began to gain traction within the scientific and management paradigm of the time, the Columbia River had already been purposed for the large-scale development of hydropower (White, 1994). In the Western landscape, preservationist goals were only applied in specific locations, or "preserves," while the rest of nature was viewed in terms of economic value and improvement (Alagona, 2013). This meant that although preservationist values were present they did not always influence early ecological management and the science it relied on. This is especially illustrated in the large-scale changes that took place throughout the western US during the first half of the twentieth century, a time when many ecological ideas were still emerging in science. For example, the ideology of maximum economic benefit was nurtured early on in resource agencies such as the USFS and the Bureau of Reclamation (Langston, 2003; Worster, 1987). This required a simplification of natural complexity in order to arrive at a more politically popular solution for the day (Hirt & Sowards, 2012).

Yet it is important to remember that river developers in the early twentieth century were able to concentrate on large dams and hydropower production partly *because* the growing field of ecology promised answers and technological fixes to the environmental degradation that would result (Fiege, 1999b; Taylor, 1999). Ecological thinking was, therefore, not always coupled with environmental values (Bocking, 2004). Early ecology was often used to address the societal goals of economic development, and people sought ways to have both a fishery *and* a heavily developed river basin. The science that is produced during a particular time is often the science that is deemed useful in fulfilling particular ideological paradigms.

Hydropower effectively divided the Columbia River Basin into an industrial river in the service of hydropower, flood control, and irrigation in the upstream reaches, leaving fisheries production only for the very lower river main stem (Allen, 2003). This spatial disparity has deeply affected ecological restoration in the basin. It has also had profound implications for ecological and environmental justice by effectively blocking and depleting salmon runs for upriver tribes. Before the 1850s, 88 percent of returning salmon were heading upstream to areas above the Bonneville Dam to spawn, but that number has now been halved to around 44 percent (Allen, 2003).

While the presence of the hydropower dams physically blocks access to upstream spawning areas, the spatial disparity that has resulted is not solely because of the development of these structures: it is also due to the technoscientific solutions that were applied to the problem, some of which Holmes himself became enrolled in. At the time the dams were built, fisheries biologists recognized the importance of habitat for fish and realized that blocking the river could potentially devastate fish runs. Yet, although habitat was a consideration, habitat restoration as we recognize it today is a more recent development. When the dams were built, habitat—especially in the headwater regions—was not recognized as a priority for mitigation funding. Instead, the main stem and lower sub-basins of the Columbia River were the major beneficiaries of mitigation dollars in the form of money for hatchery development. Ironically, as the impacts from climate change are increasingly felt, the headwaters are being called upon to provide habitat in the form of high-quality cold water refugia. And, even though mitigation from hatcheries focused on fish production in the lower Columbia River, it wasn't until fairly recently that the estuary was even considered eligible for mitigation funds for improving or restoring habitat, because habitat below the dams was not considered to be affected by the dams themselves.

In 1938, the Mitchell Act was the first major legislation aimed at addressing salmon decline due to dams. The act is emblematic of the way that the river was imagined at the time—as the techno-natural or "organic" machine—with technological fixes as the main solution (White, 1994). Although legislation such as the Mitchell Act included language to improve stream habitat and even to survey the upper tributaries, the main funding package went toward building, sustaining, and researching hatcheries and hatchery production. Meanwhile, the mitigation money that was allocated to habitat went toward building fish ladders, developing screens for irrigation canals, and, rarely, to addressing pollution (Allen, 2003). In the end, only 5 percent of Mitchell Act funds went toward this type of "habitat" work (Allen, 2003). Mitigation actions for the damage that dams caused were also implemented in 1946, in the form of the Lower Columbia River Fisheries Development Program and John Day Fisheries Mitigation, yet both of these interventions focused almost solely on hatcheries. The program eventually added a habitat initiative, which led to stream-clearing, constructing bypass around dams, and screening irrigation ditches. But all of this was too little, too late, as headwater runs had already been decimated from their previous abundance (Allen, 2003). Today, these types of fish passage projects, while improving habitat access, would not generally be considered "habitat restoration."

RESTORATION: PROCESS-BASED VS. ENGINEERING-BASED

Debates that center around restoration goals and purpose continue in stream restoration today and often bubble up between those who on one side ascribe to a more engineering-based restoration and those who on the other prefer a more process-based approach. Although this dichotomy is true in many places, the division is especially apparent when it comes to stream restoration in the arid US West, which has been so defined by river alteration to meet certain management ends such as irrigation or hydropower (Worster, 1985). Salmon habitat restoration takes many forms: riparian, river or stream, and wetland restoration, which all have many styles and iterations[1]. As I will describe throughout the following chapters, restoring salmon populations to the Columbia River Basin means all of these things and more; and, since the Columbia River Basin is a highly regulated river system, many reaches are not being restored to their full ecological function, but simply being made potentially more tolerable or survivable for salmon by whatever means possible—practically and politically.

River and stream restoration, as a subfield of ecological restoration, includes in-stream restoration, such as modifying river channels or introducing in-stream structures, but it also includes changing the interaction between wetlands and riparian areas in order to improve ecological processes that have been degraded (Wohl et al., 2005; 2015). Projects today often include a combination of these tactics in order to rehabilitate rivers, but the desired, eventual goals can be very different: restoring a river to behave in a particular way in its streambed is a much different goal than restoring floodplain connectivity. These goals are often constrained by property use and ownership. For instance, a floodplain cannot easily be reconnected to its channel if it contains homes, roads, or farms. These constraints are one reason why engineering-based methods continue to be used today, but providing some background on the differences will be useful for understanding why the controversy between the two methods of restoration, process-based and engineering-based, endures.

In the 1930s and before, early river restoration projects in the United States, especially on the East Coast and in the Midwest, were often aimed at creating better opportunities for recreational fishing (Thompson, 2006; Thompson & Stull, 2002). These projects took a structural, engineering approach to altering the channels of streams by adding engineered, in-stream structures that consist of dams and deflectors made out of large wood or rock anchored to the banks with cables (Thompson & Stull, 2002). Meanwhile, in the Pacific Northwest, large woody debris was being removed from streams in order to facilitate the passage of migratory species, such as salmon. There were little to no scientific studies done to understand whether or not in-stream structures or large woody debris was beneficial to fish species (Thompson and Stull, 2002). Regardless, the methods for restoration up until the late 1990s (and even today) were heavily focused on altering the form of the river channel itself. These methods are what are now known as engineering-based approaches (Thompson and Stull, 2002; Rosgen, 1994).

They can be contrasted with process-based approaches, which focus on floodplain connectivity and prioritizing river function in terms of ecological processes (Kondolf et al., 2006). Process-based approaches can be as simple as reconnecting rivers by removing artificial barriers such as culverts or dams, or they can aim to reconfigure the river completely, reshaping the stream and the adjacent riparian areas (Bernhardt & Palmer, 2011). The idea is to let natural processes return: to let nature do the work. "Process-based restoration aims to reestablish normative rates and magnitudes of physical, chemical, and biological processes that sustain river and floodplain ecosystems" (Beechie et al., 2010). Although restorationists in the region had been talking about the process-based approach for restoration of salmon habitat since the 1990s, it wasn't until the paper by Beechie and coauthors, "Process-Based Principles for Restoring River Ecosystems" (Beechie et al., 2010) that process-based restoration really began to take off. According to Tim Beechie:

> It didn't really all come together for us until the 2010 paper. We'd been giving the presentation, and I tried writing the paper a few times, but it just took a long time—letting the paper simmer—and finally we figured out what we really needed to say.

After the publication of their article, the method really began to take off in the Pacific Northwest, and it was quickly adopted by the Northwest Power and Conservation Council as the standard approach to restoration for its funded projects. As the following chapters demonstrate, however, in practice many of the engineering-based approaches continue to be used, often underpinned or in conjunction with the process-based philosophy.

The debate between engineering and process-based restoration exposes how the origin stories of the field of restoration continue to influence the approaches and goals that restorationists employ even today. While the school of thought that evolved from Leopold and others in the midwestern US aimed at restoring ecological processes to degraded lands, a more engineering-based style of restoration developed in western government agencies and land grant institutions such as the Civilian Conservation Corps and the Army Corps of Engineers (Thompson and Stull, 2002). This engineering-based perspective views terrestrial restoration as a way to reclaim highly degraded land, usually in places such as mine-tailing sites where all ecological processes either have been completely altered or no longer exist (Allison, 2012). Often these landscapes need complete reclamation. Restorationists facing this kind of degradation sometimes opt to use a pragmatic approach with the goal of establishing "any ecosystem" instead of none at all (Allison, 2012). While these diverse strands of restoration thought somewhat merged in the founding of the Society for Ecological Restoration in 1988—where restoration of ecological processes became the goal of all types of restoration—in practice the two are still often at odds. Despite the field coalescing around influential articles and statements from leaders in the field, the debates are ongoing—especially in the field of river restoration (Ehrenfeld, 2000).

As restorationists at the River Restoration Northwest Symposium learned about the new paradigms for river classification and the concept of Stage 0, they dealt with the emerging concepts in the ways that an epistemic community does: through rigorous experimentation, contestation, and debate. Many ask whether or not restoration is beholden to returning a site to a specific, historical ecosystem or instead, if its aim can be to achieve a state in which ecological processes are reintroduced by rebuilding ecosystem structure and function. This is certainly still a debate within the subfield of river and riparian restoration: where mandates to restore ecosystem processes must be balanced with flood protection for infrastructure, as well as demands for flow regulation for hydropower. In the Columbia River Basin, where so much of the river has been drastically transformed over the course of the twentieth century, it is difficult, if not impossible, to separate out the "ecological" from the "engineered" river that exists today.

One speaker at the conference was not convinced that engineering in-stream structures would be able to counter the large-scale and potentially drastic effects of climate. "It's like filling in the potholes of the street that is climate change," he said. Instead, he advocated the methods prescribed in Beechie et al.'s (2010) influential article, which require actions "at a scale commensurate with environmental problems" (p. 209). These process-based approaches are based on boosting stream complexity by connecting it to the floodplain. Many people who ascribe to the method even speak in terms that give the river itself agency: "let the river be what it wants to be: multiply threaded," said one conference participant. Or similarly, another speaker was adamant that through "passive restoration" the river will "tell us which way it wants to evolve." For many of the more engineering-based restorationists in the room, including those who work in valleys constrained by development and infrastructure such as roads and homes, this kind of talk can be tough to take.

But, nevertheless, a new restoration paradigm was being discussed. As one speaker pointed out to the conference audience, this label, "Stage 0," and the new object of study that it was making visible, could help lead to the social acceptance of braided floodplains themselves. These types of changes in the theory, the practices, and the metrics used to understand new scientific objects are constantly rippling through epistemic communities. They not only affect scientific understandings and discussions at conferences, but they also affect the way that the environment is managed and valued. As I will show in the following chapters, some of these shifts are now emerging in the field of restoration as restorationists adjust their practices and concepts to deal with climate change. So, given these challenges, how did the shift to a focus on restoring habitat (as opposed to other methods of recovery such as hatcheries) occur, and how did the current science of restoration in the basin emerge?

EMERGENCE OF COORDINATED HABITAT RESTORATION IN THE COLUMBIA RIVER BASIN

During the 1930s and 1940s, stream restoration was mainly focused on in-channel "improvements" (White, 1994). People believed that large woody debris should be removed from streams because it blocked fish access, a practice that is now regarded as counterproductive to improving salmon habitat. Meanwhile, the USFS and other agencies were also conducting some work in headwater areas in the Pacific Northwest in order to manage streambank erosion through stabilization (White, 1994). The work that Holmes was doing during this period—tagging fish and trying to understand their life cycles, measuring the effects of the dams, and surveying their historic territory and run sizes in the upper tributaries—was shifted to focus on finding ways to mitigate the damage that the dams had caused by creating passage around them. While fish passage became the solution for mitigation at that time, that mitigation could have taken many different forms. This choice to mitigate through creating bypass structures and hatcheries has henceforth shaped the science of the region.

As we saw, Holmes would come to spend his later career engineering fish bypass for Bonneville and other dams, still one of the main tactics for mitigating the effects of habitat loss. While it could be argued that these mitigation efforts are not comprehensive enough to "push the dial" in salmon recovery, the mitigation efforts still in place today, and the science that has followed suit, are nonetheless partly due to the passage of some of the most important and far-reaching environmental legislation in the world. The laws that were passed during the 1970s—the National Environmental Policy Act (NEPA), the Clean Water Act, and especially the ESA—continue to influence the science that is done in the Columbia River Basin in important ways. As chapter 1 described and the rest of this book will continue to illustrate, the ESA still drives the types of environmental management interventions in very particular ways: the kinds of mitigation efforts used to recover species; and the scientific work that would be needed for this, including metrics and practices that were necessary to understand how mitigation actions would meet recovery goals. As I have shown through the example of the biological opinions in the previous chapter, it is not a stretch to say that most salmon habitat restoration that occurs within the Columbia River Basin can be directly attributed to the ESA.

This period of environmental regulation also saw a resurgence of tribal political power, as tribes throughout the basin reasserted their treaty rights. In 1855, the Nez Perce, Umatilla, Warm Springs and Yakama Tribes signed treaties with the US government, reserving the right to hunt fish at "usual and accustomed" sites. The Shoshone-Bannock Tribes as well as other upriver tribes also had their own treaties, guaranteeing them the right to hunt (and fish) across their territories. Yet, many of these customary fishing sites were destroyed when construction of the dams inundated them with reservoir waters, thereby impinging on their treaty rights. Through a concerted effort and long-fought civil disobedience and resistance by the tribes, in 1969, *US v. Oregon* (302 F Supp. 899) designated tribal fishing sites along the Columbia and guaranteed tribal fishers 50 percent of the overall harvest. This development made recovery of salmon even more critical as tribes now have greater rights to demand mitigation and recovery, and environmental justice has finally been put on the legal table.[2]

During the 1970s the conflict between hydropower and fish was therefore ratcheting up, and in 1980 the Northwest Power and Planning Act was passed with the vision of creating a plan that would maintain energy production and mitigate damage to fish and wildlife. Out of this act, the NW Council was formed, and the council would become one of the main agencies driving restoration of salmon and their habitat by recommending where Bonneville Power Administration mitigation money should be spent. While the Northwest Power and Planning Act anticipated that the NW Council would call for actions to mitigate the damage to fish populations from the dams, the NW Council also assumed that because the dams completely block large areas of historic habitat, this mitigation would be partial, since wholesale mitigation would likely necessitate removal of the dams. Because of this assumption by key players, much of the mitigation for the habitat that is completely blocked is done in offsite locations, not in previously accessible habitat. In order to establish which areas should be improved or restored, NW Council began restoration prioritization efforts in the 1980s, as high-quality spawning and rearing areas began to be identified. The first such Columbia River Fish and Wildlife Program was produced in 1982, and habitat improvement was a critical element of the mitigation plan (NW Council, 1982). The ESA requirement for fish and wildlife agencies to develop BiOps—as discussed in the previous chapter—has driven the science that is used to understand the effects of dams, but BiOps have also helped establish the importance of habitat restoration as an important mitigation tactic for offsetting dams' damaging effects.

While environmental impacts from dams were driving a need to protect and restore salmon habitat, much of the early riparian habitat monitoring, protection, and restoration that took place in the Pacific Northwest was originally done in the field of forestry, out of a concern for the impacts of logging. In 1987, *Streamside Management*, published by the College of Forest Resources at the University of Washington, became one of the first how-to guides for riparian restoration in the region. During this period, some of the first interdisciplinary symposia aimed at riparian restoration science were also held. This development of restoration within the field of forestry was partly driven by the regulatory framework of the Northwest Forest Plan, which aimed to make logging practices more accountable to environmental damage. In the early 1990s, an interdisciplinary cross-pollination at the University of Washington began to be cultivated. Researchers from forestry and fisheries departments, as well as engineering, formed a working group called the Center for Streamside Studies. A main concern on the research agenda was degradation to water quality through impacts from logging roads and the fine sediments they let loose. Dead wood and large woody debris in streams also were recognized as important and began to be studied through wood surveys and pool area studies. While large woody debris in streams had been discussed in the literature since 1979 (Keller & Swanson, 1979), it only became a riparian restoration tool in the early 1990s (Naiman, Beechie, & Benda, 1992).

These studies helped inform early riparian restoration practices in the region and set into action a trend toward "process-based" restoration that began in the 1990s, as landscape and watershed scale approaches were also introduced (White, 1996). Although throughout the 1990s the focus for many restoration agencies remained on placing fish screens in irrigation ditches and fish passage to blocked areas, researchers like Tim Beechie and George Pess were helping to shift the focus from engineering-based to process-based restoration. Throughout the 1990s, scientists at NOAA Fisheries, the University of Washington, and other agencies and universities wrote about and studied the benefits of process-based restoration and the use of remote sensing to prioritize and monitor restoration success, as well as the need for large woody debris in streams. Research like this helped shift restoration thinking toward creating and improving riparian habitat through restoring ecological processes and reconnecting streams to floodplains—and it was an exciting and innovative time for riparian restoration in the region.

At the same time, Columbia River Tribes were also pushing for a change in focus for ecological restoration in the region. In 1995, the Columbia River Inter-Tribal Fish Commission published *Spirit of the Salmon: Wy-Kan-Ush-Mi Wa-Kish-Wit* (CRITFC, 1995), a tribally driven restoration plan that pushed for comprehensive habitat restoration as a key recovery strategy through the development of sub-basin plans. During this same period, salmonids throughout the Columbia River Basin were being listed under the ESA, in large part due to the efforts of tribes. By the mid 1990s, recovery planning was in full swing. The ESA requires monitoring every five years, so large-scale monitoring programs had to be put in place. Recovery plans, including baselines and benchmarks for populations and habitat areas had to be developed and updated using the "best available science," calling on ecologists, geomorphologists, and biologists throughout the basin to get to work monitoring and understanding watersheds across the basin.

In 1996, the NW Council created the Independent Scientific Review Panel (ISRP), which would serve to independently review all projects funded by the BPA. When they met in 1996, the panel recognized the overwhelming skew toward mitigation through hatcheries and the lack of focus on habitat restoration. Despite this, in 1996, a National Research Council report titled *Upstream* was published. It called for the "rehabilitation" of salmon populations, but rehabilitation mainly relying on hatchery technology, and included only long-term visions of restoration that may eventually occur, someday. Then in 2006, the ISRP finally published *Return to the River,* which refocused the recovery effort on process-based restoration. While language and funding still favored technological fixes, a shift in restoration practice had occurred. *Return to the River* introduced an ecosystem-based approach and described what it would mean to mitigate habitat loss and recover salmon through a focus on habitat restoration (Williams, 2006). This scientific work, and the gradual shift to embrace ecosystem-based, and eventually process-based, approaches in turn provided a conceptual framework for the Fish and Wildlife Program of the NW Council, as well as for the recovery plans developed in the coming decades. Ecological restoration had finally arrived in the Columbia River Basin. Yet the best way to restore salmon to the river was far from decided, as restorationists themselves were still actively debating and defining the field.

CONTINUING DEBATES WITHIN THE FIELD OF RESTORATION ECOLOGY

In 1987, ecologist Anthony Bradshaw famously asserted that ecological restoration was the "acid test for ecology," believing that it should form the experimental basis for the discipline of ecology itself. Bradshaw (1987) recognized early on the difficult relationship between management and science in ecological restoration, and he worried that restoration's roots in ecology were getting "left behind" as it turned to the tools of engineering and natural resource management to transform nature. In an early collection of essays, restorationists pondered the relationship between science and ecology and worked to define the burgeoning field by debating—and drawing boundaries around—what restoration was and was not (Jordan, Gilpin, & Aber, 1987). As the field has emerged from multiple disciplines and geographical locations, restorationists, ecologists, and philosophers have vigorously debated the purpose, goals, values, and meaning of ecological restoration and its relationship to management, science, and the public. These debates have been formative to the field and they are ongoing.

Throughout the development of the field, there have been voices that have warned about "dangers" that could derail the restoration enterprise entirely. These warnings include fears of creating an overly technological restoration discipline that would have unintended, negative effects on society should humans become convinced they could "fix" any ecological damage they cause (Higgs, 2003). Some authors have even gone as far as to call restoration a "big lie" because it cannot meet our expectations to fix what humans have destroyed, and instead advocate for preservation as the main focus of conservation efforts (Katz, 1992). Yet ecological restoration is often seen as a purely practical matter, where science and technology can offer solutions to predetermined problems (Nilsson & Aradóttir, 2013).

These debates have murmured on through the years, making clear the political and social nature of ecological restoration (Light & Higgs, 1996). For example, one of the recurring discussions concerns whether or not cultural values should be included when considering restoration planning (Higgs, 1994). Eric Higgs (1994) was concerned that an overly "technical" restoration discipline could come to dominate the field and ignore local community participation and involvement in restoration work. This dispute circles around the problematic division between society and nature in much of environmental management literature, and some see participatory restoration as a partial solution to this problem (Higgs, 1994; Light & Higgs, 1996). While many of these efforts have been successful in widening the participation in ecological restoration, like many transdisciplinary, citizen science, or public engagements with science, they have also had shortcomings in terms of both the results, and the engagement and communication from both scientists and the public (Eden &Tunstall, 2006; Tadaki & Sinner, 2004). A coincident worry centers around restoration becoming overly commodified and commercial as more and more private restoration enterprises seek to profit from the large amounts of money being spent to restore degraded landscapes: thereby potentially forfeiting not only restoration's scientific basis, but its democratic potential (Lave, 2012; Lave, Doyle, & Robertson, 2010; Light & Higgs, 1996).

The role of science itself in restoration has also been a subject of ongoing debate. Higgs (2005) wonders whether or not there is a danger of a "scientific authoritarianism" taking over the field and losing sight of other forms of knowledge. While another restorationist, Robert Cabin (2007b), questions whether science is a useful framework for ecological restoration in the first place, arguing that the complexity of nature and land management is fundamentally mismatched with the "culture" of science. Instead, he offers a framework for a "trial-and-error" restoration practice or, borrowing from Leopold, an "intelligent tinkering" approach as opposed to a "more rigorous, data-intensive scientific methodology" (Cabin, 2007b, 2011). The "intelligent tinkering" model of restoration practice that he outlines blends what he calls the "attributes of good science (e.g., objectivity, hypothesis testing, and rigor) with attributes of good practice (e.g., technical skill, local knowledge, relentless passion)" (Cabin, 2011). He sees value in the legitimacy and financial benefits that a scientific framework can provide—and thinks it is important to consider when "formal science" may be appropriate—but he also wants to acknowledge when it might not be as helpful in forwarding the goals of restoration (Cabin, 2007b). While (Cabin, 2007b) believes that there is a need for more professional support and reward for engaging in this kind of "tinkering" and experimentation, he points out that the goals of scientific experimentation can often be at odds with the need to "get things done on the ground" (p. 2). Cabin's (2007) ideas, although seemingly innocuous to an outsider, provoked a fierce debate within the journal *Restoration Ecology*, with lively exchanges about the role of science in restoration, the definition of science itself, and whether or not the long time frame science requires to demonstrate results in restoration is actually the real problem for restoration as a field (see Cabin, 2007b, 2007a; Giardina, 2007).

While some restorationists see debates such as these as a fundamental science-practice divide in restoration, others view them as reflecting a problem of science "driving" restoration. Yet a closer look at the values behind restoration goals and priorities demonstrates that this is not the case. These tensions and debates about the fundamental drivers and goals are things that all ecologists (and scientists) struggle with (even if not openly) when setting up an experiment in the field or in the lab. As they choose the right time scale, decide how to deal with uncertainty, and factor in the limits of knowledge, they are making judgments that necessarily incorporate non-epistemic values (Douglas, 2000). This is not due to a difference between "the rest of science" and restoration but instead from dynamics imbedded in the practice of scientific work more generally (Douglas, 2000). While these non-epistemic goals and values are present in all restoration efforts, the examples in this book will provide details in one iteration of this expression of science and value in natural resource management. As I describe throughout this case, there are many drivers in the socio-ecological-technical system that is the Columbia River Basin today, and they all have feedbacks that both push and pull. In my research I found that, instead of forcing restoration into definitional boxes, restorationists in the Columbia River Basin are reorienting themselves to solve the practical, "on-the-ground" problem of restoring salmon and their habitat in a changing climate by using creative, adaptive, and interdisciplinary ways; these are examples of adaptation in science.

CONCLUSION

As I explored in chapter 1, ecological restoration in the Columbia River Basin is supported by specific institutions and organizations, including large, basin-wide organizations and entities such as NOAA Fisheries, CRITFC, the NW Council, and BPA, as well as smaller local and regional watershed councils, individual tribal and state natural resource agencies. Meanwhile, the legal infrastructure and norms of environmental law drive salmon recovery and restoration of critical habitat. The most influential of these is the ESA, although other environmental and natural resource laws, such as NEPA, the Northwest Forest Plan, and Tribal Treaties; agreements such as the Fish Accords; and water quality standards are also important. Organizations that influence science include fisheries and forestry departments at universities, tribal research departments, and research consulting firms. Ecological restoration, in turn, has developed in relation to these. In many ways, the ESA, in particular, launched ecological restoration in the Columbia River Basin, and many of the institutions and organizations in place today owe their creation to this law. As one restorationist put it, the listing of salmonids in the basin "kicked things into a different level and into a higher gear."

Yet despite the mandates and the money, restoring salmon and their habitat to the Columbia River Basin is still a herculean task. In a sense, people in the Pacific Northwest are trying to do something that has never been done before: to maintain a highly regulated river system that supports a hydro-industrial complex while at the same time maintaining viable anadromous fish populations. Restoring salmon to the Columbia River Basin—"fixing" all of the problems that have been caused by an industrial hydrosystem—is, in a sense, a kind of mega-experiment on a massive scale. Yet this large-scale experiment was never meant to be one, and as such, there is no overarching experimental design. Instead, the restoration effort has evolved and the epistemic community of restoration specialists has adapted as new problems and information has arisen, or society demands new forms of knowledge.

Looking back to Holmes's work in the 1920s, although still recognizable, relevant, and even useful to scientists today, it was also different in important ways. Scientific practice is conducted within the context of the cultural values and norms of a specific time and place. The sociocultural and environmental imaginary of the Columbia River Basin and the salmon within it: the objects of Holmes's scientific work, were different scientific objects than what they are today. This is not because the fish or the river or the water are fundamentally different; it is because the science that exists today is situated in a very different time and place and it is driven by different values and societal needs—it relates to its object differently. New ways to manage nature emerge from scientific practice, but they are also created: institutions require different kinds of knowledge in order to make decisions. As I have shown, one of the main institutions driving science in the basin is the ESA, but there are many others that also influence the science in the basin, including the epistemic community of restorationists. As knowledge about the river is produced, the ways in which we view these scientific objects will change. Although the technologies that Holmes employed were different—the instruments and facilities that he developed—the institutions and organizations that supported and constrained his work have also evolved.

Restoration in the Columbia River Basin is, in a sense, a "large-scale experiment" (Gross, 2010). Our society is only just discovering what it means to manage such a task. This experiment is, in many ways, very different from the one that Holmes developed throughout his career as he traveled around the basin trying to understand salmon and their life cycles. Yet in other ways, his practice seems familiar to the scientific work that is being carried out in the basin today. Holmes' early work was concerned with measurement, with tracing the migrations of salmon, and generally understanding and classifying the species and runs. Yet he was also concerned with understanding the broader ecological processes within the Columbia River Basin, and he was particularly interested in establishing a baseline understanding of where salmon habitat was located. His work was similar, in many ways, to the habitat monitoring work that is ongoing today—he was counting, making visible, and giving a name to salmon and their habitats and life cycles. Yet, instead of being able to measure and monitor habitat in the way that a restorationist may like to today, Holmes did the best he could with a highly qualitative description of what kinds of habitats salmon seemed to like and where their habitat had been lost. This was not due to his lack of precision or rigor as a scientist, but due to the kind of work that was being funded at the time and the scientific methods and metrics of ecological restoration that were available to him.

As restorationists today cope with the onset of climate change, they too will work with what they are given, developing and adapting their practices where they can. This process can even be creative as they employ both conventional, engineering-based methods of restoration, and as they experiment with process-based restoration on a scale that has not been done before. The contingent nature of scientific work is an ongoing struggle for those working in the field. There is no overarching experimental design for restoring salmon habitat in the Columbia River Basin, and the results of this large-scale experiment are yet to be seen. Yet, one thing that we can observe is what restorationists are doing today in their practices and in their work. While the previous chapters outlined a broader background to the historical and institutional framework to set out a basis for understanding restoration science in the basin, I now want to shift the view to individual practices. Individual restorationists are doing the daily work of restoration itself, and this is where adaptations to climate change can be found. The next chapter, therefore, "zooms in" to individual restorationists and their work, in order to describe both the mundane tasks and the creative insights that go into a restoration career during this time of climate change.

3

The Work of Restoration in a Changing Climate

IT WILL HOPEFULLY be becoming clear by now that the work of ecological restoration involves many different things. The activities that make up restoration, the people that conduct this work, the places where it occurs, and the styles of restoration themselves are diverse. A comprehensive catalogue of what goes on in the Columbia River Basin in terms of restoration is therefore not the point of this book. Instead, I've taken a different tactic—I want to both "zoom in" to the detailed scientific work of individual restorationists, their concerns and their hopes, but also "zoom out," taking a broader-scale look at what is occurring at the regional scale. As the previous chapters explained, this includes the institutions, the histories, and the policies, that co-produce science along with the work of the people involved.

However, an important dynamic in this narrative draws us even further out in scale: the global changes in climate. How is the global phenomenon that is climate change altering the behaviors and decisions of those working on restoration? How is climate change affecting their ability to do their work? How is it changing what they do and what they think? This chapter describes the work of two restorationists—an aquatic ecologist and his field crew leader, a fisheries biologist, who have worked on a large-scale and transformative project over the course of fifteen years. This narrative allows me to introduce some of the specific things that restorationists do, as well as the ways in which climate change affects restorationists and their work, both practically and affectively, on an individual level. Their efforts to restore a watershed also illustrates the issue of "shifting baselines," in which what is assumed to be "normal" in terms of ecology can shift over time, with each generation of scientists accepting a more degraded environment as a starting point and "natural" (Pauly, 1995). This is a problem by no means unique to restoration—or to the Columbia River Basin. It is relevant globally, as environmental change occurs in ecosystems around the world.

Sean Gallagher literally stumbled into the river and into his restoration career almost by accident. In 2003, he was up to his knees in Tarboo Creek, which at that point resembled more of a ditch, in his backyard at the bottom of a valley located on the Olympic Peninsula of Washington State[1]. As the story goes, Peter Bahls, the director of the Northwest Watershed Institute (NWI) and an aquatic ecologist and conservation biologist, drove over a bridge and noticed Gallagher mucking about down in the creek. Asking him what he was doing down there, Gallagher said he was just poking around for fun. "Well, if you think that's fun, I'll give you a job," Bahls replied. A dozen years and miles of stream restoration later, Bahls and Gallagher were still working for NWI. In the meantime, Gallagher had earned himself a bachelor's degree in fisheries biology and both restorationists have poured their efforts into the physical labor of transforming the river, gaining an intimate knowledge of Tarboo Creek and Dabob Bay, the estuary downstream.

Bahls's vision for the restoration of the Tarboo Creek–Dabob bay watershed was (and still is) far-reaching. He wants to restore or protect over seven miles of stream—basically the entirety of the Tarboo Valley, as well as the Dabob Bay estuary at its mouth. And, he has been successful in this effort. To date, over 4,000 acres of land have been conserved in cooperation with many project partners and landowners and over 300 acres and four miles of valley bottom streams and shorelines have been restored.

The Tarboo Valley had been heavily degraded from clearing for agriculture in the late 1800s, and the creek mostly ditched and its wetlands drained by the early part of the 1900s. Yet, despite this, it is still important habitat for populations of chum as well as coho, and the estuary is still one of the most pristine in the region (Freeman, 2018). While Bahls coordinated this landscape-scale restoration effort from both his office and the creek, Gallagher was usually found out in the field on seasonal work, helping with surveys, and supervising the winter or summer field crews in planting trees, clearing invasive species such as Scotch broom and Himalayan blackberry, or removing derelict buildings, junk cars, and whatever else had accumulated in the valley over the course of a hundred years of rural development—and rural decay. Over the past seventeen years, NWI, along with many other partners including local tribal governments, natural resource agencies, as well as schools and nonprofits, transformed the stream from a degraded ditch clogged with invasive reed canarygrass into a salmon-bearing stream. It is now well on its way to becoming more than just the vision of a restored forested valley found only in the dreams of restorationists and salmon: walking through the valley in 2019, you can sense the beginnings of what will someday, hopefully, be a Sitka spruce–dominated wetland forest.

Gallagher and Bahls are both understated guys. With his unpretentious, old-school skater-punk vibe, you could easily miss the fact that Gallagher has many covert talents. Besides being a fisheries biologist, he is an acclaimed artist and carver who draws on his Iñupiaq King Island Alaskan heritage for inspiration. He has also traveled the world teaching people how to build the unique style of sealskin kayak that hails from King Island itself.

Bahls's dedication to salmon comes naturally. He began fishing with his grandparents when he was twelve and has not missed a summer steelhead season in the Columbia River Basin for the past forty-four years. With his laser focus on salmon, you would never guess that Bahls started his work as a biologist doing everything from studying ants in Panama, to tracking brown bears in Alaska, to conducting one of the largest ecological surveys of mountain lakes in the western United States for his master's in fisheries from Oregon State University. After working for the Port Gamble S'Klallam Tribe as a habitat biologist and then as a senior level fishery biologist for a large consulting firm, he and his wife Jude

Rubin, NWI's stewardship director and a botanist, founded Northwest Watershed Institute to restore the Tarboo Creek watershed.

In his own unique and laid-back way, Gallagher agreed to take me on a tour of the valley, walking me through how all of this restoration was *physically* accomplished; in addition he told me how the stream came to be known in a scientific, as well as a tacit, way through his own work.[2] One thing that quickly became apparent on our tour, is that Gallagher has thought *a lot* about this valley and the work he has done here on and off over the past fourteen years. If you can get him talking, he has some stories to tell.

Gallagher started us out at the mouth of the creek. Here, Tarboo Valley meets a wide, muddy bay full of driftwood and seagulls fighting over the smells of low tide:

> My first job was to survey the creek with Peter. The whole watershed. Before we did any restoration. We wanted to see how much habitat there was, to see what the project would be. We found out what habitat limitations for the fish there were, and one of the big things was that it was blocked and redirected into ditches. There used to be a lot more fish here before those changes. A lot of what we were doing was just making room for more fish.

Surveying the stream involved splitting the creek up into sections, all the way from the mouth to the small lake that feeds it. Each section was measured in detail. The goal was not only to take account of the degradation, but also to create a baseline from which to measure change—a starting point for the restoration project from which they could quantify success (or failure). The depth and width of the stream was measured in cross-sections at various points for the entire 7.5-mile length. Gallagher recalled, "We'd walk up the stream and survey it, sections at a time. It would actually take quite a while!"

Flow velocity was noted. The type of sediment was classified. Vegetation, invertebrates, fish, and other species were counted. The amount of shade was measured, and large woody debris—or LWD as it is known in the restoration community—was counted. Now known to be an important and beneficial aspect of fish habitat, LWD used to be thought of as a barrier to salmon passage (Naiman, Beechie, & Benda, 1992).

Although Gallagher's and Bahls's job was technically to measure, monitor, and restore the stream, a large part of their task was also to interface with the community in the valley—to explain to people what they were doing and why. Restorationists are out in the world, doing things and changing things, and so a large part of their work is putting a public face on science. Big changes were taking place in the valley. It was being transformed in a way that to many of the old-timers seemed anachronistic to the way they thought about land improvement—draining and clearing fields, or channeling the stream into a straight ditch to better drain wetland pastures. Yet Bahls was sensitive to this issue, taking community perspectives into account from the beginning of the project:

> As part of the initial watershed assessment, I did a lot of historical research—interviewed the elders who grew up here, examined old photos, as well as aerial photos, read the survey notes from the first surveyors for the General Land Office back in the 1870s, and found relict stream areas that had not been so heavily altered. All this info gave us clues as to what the stream system was like.

Gallagher took me further up the valley, to an area that used to be dairy farm but had been abandoned since the 1960s. Although a lot of the outbuildings and junked cars had been removed, NWI decided to leave the old, iconic dairy barn as a reminder of the valley's heritage and as a permanent base for their restoration crew and supplies. The barn's eves were coated in cliff swallow's nests, and as we entered, we disturbed a barn owl. Gallagher greeted the bird as it ghosted overhead: "Hey, barn owl. Sorry, dude."

Looking out at the valley and the old farm fields, it was difficult to for me to tell that thousands of hours of work over the past seventeen years had gone into creating and maintaining the scrubby forest that was springing up. Over 90,000 trees had been planted in the valley. Walking through the plantings, which were about shoulder-height, Gallagher pointed out the different species that they had carefully arranged and tended:

"Ninebark . . . spirea. . . . We planted lots of spruce—historic photos showed that the valley used to be full of spruce." He was clearly proud of the success of the healthy, young trees; although there have also been some setbacks. Many of the spruce have been suffering from an infestation of spruce weevil. Regardless, many of the trees are doing well:

> When we started here it was just a field. This was all planted about six years ago. I haven't seen these trees in a while, some over there didn't do so well, we had to take the blackberries out about ten times. But, looks like some are growing. It's all in stages.

These plantings, which cover over 300 acres along much of Tarboo Creek, are only part of the work that has gone into restoring the valley. Much of Tarboo Creek had been turned into a ditch in order to drain the fields and make them available to grazing, a project which had only been partially successful, resulting in sub-standard farmland. In order to create salmon habitat, the stream needed to have meanders reintroduced: to be "re-meandered," and otherwise complexified, and released from its incised banks into the floodplain and the new forest. This would create overwintering, rearing habitat for salmon smolt and reengage the floodplain. In addition to re-meandering the stream, to create additional habitat hundreds of tons of LWD was brought in and placed in locations hydrologically calculated to help increase complexity. Gallagher showed me one of the re-meanders that had been put in, but the stream's course had completely changed and gone in a different direction than the engineering plans had calculated: "the swale turned into the main stem, which was kind of crazy." But, according to him, this was no big deal: "it just does what it wants."

"We aren't restoring the stream per se," said Bahls. "We are trying to restore a functioning, healthy floodplain system. In this case, that means beaver that have moved in and are adding all sorts of ponds that are primo rearing habitat for juvenile coho salmon and creating a whole network of channels that are restoring complexity and functioning to slow and spread out potentially damaging floodwaters." Indeed, Northwest Watershed Institute's project is an exemplar of process-based restoration.

BASELINES

Our tour continued down the valley. Gallagher wanted to show me a special part of the watershed, one that was well off any trail. We parked on the side of a dirt road and set out on foot through a dark forest full of moss-covered old-growth trees. Ravens were causing a ruckus in the distance, alerting the rest of the forest inhabitants that a couple of rare, human visitors were in the neighborhood. We wound our way through large sword ferns that covered the forest floor and under vine maples that shaded the creek. Although we were in the midst of a rare spring drought, the forest still felt damp and cool. But, the size of the trees was the most striking aspect of the forest. There were alders over two feet in diameter, and many hemlock, cedar, and spruce three and four feet across. After a while, we reached our destination: an immense Sitka spruce at least six feet across. Gallagher's guess is that it might be one of the largest spruce in the region—and here it was, its roots twisting through a part of Tarboo Creek itself, an ancient grove that was now rarely visited by humans.

But, Gallagher's reason for bringing me to this spot was not only to visit this giant old friend. He wanted to show me what the valley looked like in the past—and what he and others believe it may look like in the future. This stretch of creek, where older age, naturally regenerated forest still exists, is being used as a reference—a historical analogue, an environmental baseline—for comparison. According to Bahls, it is one of the last pieces of older forest along Tarboo Creek. This is what restorationists would call a "historical reference site," and it is a common experimental and conceptual tool used to plan a restoration project. But for Gallagher, it was clearly more than that: it was not only a measuring point, but also a source of inspiration, a future imaginary for what the Tarboo Valley *could* be: "This is an inspiration, this beautiful forest. The *old* forest."

Restorationists use different kinds of baselines in different ways, making "baselining" a varied practice. Comparisons using "undisturbed" sites as reference baselines, for example, have been fundamental to experimental design in ecological science since the early 1900s (Adams, 1913). Measuring success or failure in ecological restoration not only requires setting historical baselines for populations, but also requires establishing reference habitat sites: analog ecological systems that act as controls for comparison or restoration goals. In addition to generating knowledge about the environment, restorationists use baselines to monitor the effectiveness of their actions, and they must therefore either employ existing baselines or establish their own. Yet, while baselines are critical for measuring change, they are also a necessary regulatory feature. In order to understand the success or failure of environmental management and mitigation measures, they are a requirement of many environmental laws such as the Endangered Species Act (Alagona, 2013).

THE PROBLEM OF SHIFTING BASELINES

One of the biggest problems that climate change introduces to ecological restoration, and ecological monitoring more broadly, is the issue of shifting baselines. For ecological restorationists, baseline-setting is a ubiquitous scientific practice that enables comparison across spatial and temporal extents. Such baseline comparisons using "undisturbed" sites have been crucial to designing experiments and understanding change (Adams, 1913). Baselines are critical for measuring change, but they are also necessary in order to understand whether or not a particular management intervention or mitigation measure has succeeded or failed. These measurements are not only useful, but they are also legally required. Baselines are therefore one of the key sites of co-production between science and law itself (Jasanoff, 2004).

Daniel Pauly (1995), a marine biologist, first described the inherently social nature of ecological baselines as the "shifting baselines" syndrome. He used the example of fisheries management targets, in which "each generation of fisheries scientists accepts the stock size and species composition that occurred at the beginning of their careers as the baseline (Pauly, 1995, p. 430). According to Pauly (1995), the shifting baselines problem has led to a highly distorted view of fish abundance, which does not account for much larger fish populations that existed historically. To counter this issue, he recommends taking "anecdotal," historical evidence into account to determine population numbers on a longer timescale. He highlights the danger in presenting baselines as pre-given, scientific fact, bringing attention to the social nature of shifting baselines. He points out that, because baselines are socially constructed, history can easily be erased and previously abundant fisheries forgotten, even perhaps by accident, leaving a depleted ecosystem as the accepted—or remembered—norm. Yet, through scientific practice baselines are renegotiated on an ongoing basis in many ways, including in response to changes in societal goals, the introduction of new information, and changes in environmental or ecological trends.

Restoration and recovery goals for the ESA and other regulatory measures often use historic baselines as metrics. Therefore, as a no-analogue future (and present) becomes more apparent owing to climate change, restorationists and policy-makers are struggling to determine the best way to measure success and failure in salmon recovery. In place of the traditional, pre-1900, historic baselines for the river, restorationists are beginning to employ more recent baselines in their work. Some restorationists have proposed using a "dynamic reference" that accounts for change in both reference and restoration sites (Hiers et al., 2012). These ideas have helped broaden the goal of restoration to include states other than historic baselines or analogs. For example, by conceptualizing "Anthropocene baselines," it may be possible to triage historic remnant ecosystems. For example, this is a tactic taken by some coral scientists, who are facing coral mass bleaching events (Braverman, 2018). At the same time, those working in many watersheds are recognizing the reality of irreversible ecological change in highly altered and human-dominated ecosystems such as regulated rivers like the Columbia (Kopf et al., 2015). There is a fear, however, that abandoning historic baselines for recovery and restoration goals could lead to people "gaming" the system (Ruhl & Salzman, 2011)—changing points of reference in ways that could minimize biodiversity (Kopf et al., 2015)—or succumbing to the shifting baselines problem identified by Pauly (1995). These fears were often voiced among participants in this study.

CLIMATE CHANGE IN THE COLUMBIA

In place of using traditional, historic, baselines for the river, restorationists are beginning to think of employing more recent baselines in order to understand what might be possible in terms of restoration in a changed climate. One of these new, "anticipatory" baselines is 2015, the record-warm year that brought mass die-offs of salmon to the fatally hot Columbia and its tributaries (Hirsch, 2019). According to climate models, the future does indeed look like the record-warm year of 2015. Restorationists throughout the basin are trying to understand and adjust to this change, an adaptation that is helping to create a new baseline for restoration: one that is based not on the past, but more on what can be expected in the future (Hirsch, 2019).

In 2015, the North Pacific Ocean was experiencing a sea-surface temperature anomaly that became colloquially known as "the blob." Surface temperatures in the Pacific had been consistently above normal since 2013, creating a "blob" of warm water that extended from the Bering Sea to California (Bond et al., 2015). This blob, while likened to the effects of El Niño years, was longer and more intense, and had much more severe effects on fish runs. These included low returns of ocean-run salmonids throughout the region and record low returns from 2013 to 2015 (Pacific Fishery Management Council, 2016). Unlike a typical El Niño which cycles through in a year or two, the blob continued to stick around for longer than expected and became the greatest deviation from normal since at least the 1980s, and possibly since 1900 (Bond et al., 2015). Critically, for salmon feeding in the ocean, this warm water prevented the nutrient upwelling that forms the foundation for the coastal ecosystem's support for forage fish (Gewin, 2015). But the blob not only affected the ocean, the warm ocean water also played a role in keeping temperatures at record-high levels over land (Mote et al., 2016), delivering a double punch of climate impacts to migrating salmonids. Some scientists believe that rising ocean temperatures may lead to more ocean anomalies—meaning ocean conditions like "the blob" will become the "new normal." Environmental shifts like these have caused fisheries managers to rethink their modeling and management tools; as a result, some managers have adopted predictive ecosystem-based models for decision-making rather than relying on past ecological baselines that may no longer exist (Gewin, 2015).

Tarboo Creek's salmon, even with restored habitat, were not immune to the effects of the blob, according to Bahls:

> I have been walking stretches of the creek every year for nearly two decades, doing salmon surveys that are added to the state's database. We had some great years of returns after the restoration, but

the past three seasons have been the worst I have ever seen. Not only are there very few fish coming back to spawn, but one year in particular, they were quite small in size. I speculate that it is a combined hit of chronic overfishing and the blob. If this keeps up, Tarboo Creek and Hood Canal's salmon will be extinct regardless of what habitat measures we take.

Uncertainty within and between social and ecological dynamics is only magnified when climate change is layered onto these spatially and politically complex management regimes within the basin. As more precipitation falls in winter in the form of rain rather than snow, water temperatures are increasing, summer flows are decreasing, rivers are becoming "flashier" from the onslaught of storm water, and wildfire risk and intensity is increasing year after year (Nolin, Sproles, & Brown, 2012). Researchers have found that changes in both precipitation and temperature are already impacting the hydrologic regime of the basin, and these changes are only intensifying (Dalton, Mote, & Snover, 2013; Mote et al., 2003). Climatic changes are expected to shift the spring peak flows in the Columbia Basin to earlier in the season while at the same time decreasing late-summer flows (Mote et al., 2003). Yet these snowmelt-driven summer flows are what allow many salmon populations to migrate to their headwater streams during the spawning phase of their life cycle. Meanwhile, the increasing water temperatures in lower elevations are also harming salmon survival and spawning, making healthy, cooler, higher elevation, headwater stream habitats even more critical as "climate refugia" (Mantua, Tohver, & Hamlet, 2010). It is also difficult to measure effectiveness when changes that are taking place are shifting the environmental baseline of the region itself. As climate change takes effect, these large-scale shifting baselines are becoming the "new normal."

CLIMATE IMPACTS ON THE FIELD OF RESTORATION

Ecological restoration of fish and wildlife habitat is recognized as a "global priority" and has been incorporated into the Convention for Biological Diversity (CBD) and the United Nations Environment Program (UNEP) (Aronson & Alexander, 2013). Further, climate change mitigation and adaptation policy itself increasingly relies on the field of ecological restoration as its scientific justification and base (Baker & Eckerberg, 2013). While ecological restoration is quickly becoming recognized as critical for maintaining biodiversity and mitigating the impacts of climate change, the field itself is also trying to adapt to and anticipate a climate-changed future where there is "no analogue" environment for comparing the present to the past (Williams & Jackson, 2007). Yet there is also a general recognition among restorationists that climate change is making restoration even more important, even as restoration becomes more difficult to implement successfully (McDonald, 2013). What will the goal of restoration be if restoring to a past ecological state is no longer reasonable or viable? As restorationists come to terms with the effects of future climate change, how will future natures be understood, valued, and compared to past reference states that have been used for comparison for so long?

Measuring and identifying ecological thresholds and change is already challenging for ecological restorationists (Hobbs, Higgs, & Hall, 2013). Because of this, and due to the effects of climate change, many of the field's prominent scholars have recognized the need to broaden the scope of the meaning of restoration (Clewell, 2009; Hobbs & Cramer, 2008). The tools of ecosystem equilibrium models and managing for historic range of variability are becoming increasingly anachronistic as historic conditions no longer exist or are at least no longer reasonable goals for restoration (Seastedt, Hobbs, & Suding, 2008). Instead, new concepts that move away from the idea of restoring ecosystems to a historical reference baseline are becoming increasingly common. Novel ecosystems, for instance, are ecosystems that contain species combinations that have not previously occurred within a given biome (Hobbs et al., 2006). Similarly, "hybrid ecosystems" recognize the need to orient toward a future ecological state (Choi, 2004; Hobbs et al., 2009). Some restorationists have proposed using a "dynamic reference" that accounts for changes in both reference and restoration sites (Hiers et al., 2012). Although concepts like novel ecosystems are becoming more widespread, they are not always accepted, and the implications for management are contentious. Nonetheless, these concepts have helped restoration ecologists reconceptualize the restoration of ecosystems to a state other than a historic analogue (Hirsch, 2019).

For restorationists working on the front lines of climate change in the Columbia River Basin, a sense of urgency is often coupled with a sense that there is not enough time to employ the scientific method. As van Diggelen, Grootjans, & Harris (2001) lamented in reference to ecological restoration almost two decades ago: "the feeling is often that the situation is so critical that one should act immediately and try to salvage all that can be" (p. 115). Restoration ecologists Jackson & Hobbs (2009) also worry that "we face serious risk that global change will outpace our scientific capacity to prescribe adaptive strategies, let alone implement them" (p. 568). Irus Braverman (2018) captures this pendulum of hope and despair in her ethnographic work on coral scientists. The field of coral science is rife with debates about whether or not to restore corals, preserve what is left, or even breed coral species that may be more resistant to climate change. Restoring to a projected future environment, as opposed to a historic baseline, involves predicting what that future may be; but it also involves the imagination, hope, and anticipation of an individual scientist (Braverman, 2018). These acts are performed through their scientific practice, and through their work, necessitating at least a partial conviction that what they are doing will actually make a difference. While some restorationists are looking to future baselines like 2015, many see any form of ecological restoration as being relevant and influential in the ways that it can prepare, or buffer, for a future environment, even if that environment is not the one that existed in the past.

WORKING THROUGH IT ALL

Sean Gallagher also remembers that exceptionally dry year of 2015:

> Yeah, I remember 2015 because normally we slash-and-burn a bunch of blackberries—put 'em in a pile and burn 'em—and you'd never have to worry about catching anything on fire, because we'd do it in the winter. Well, that winter we made this big fire and we were burning all these blackberries, and so we were sitting, eating lunch, talking, and the field caught on fire! That never happens. The grass was dry enough to catch fire even in the winter!

I asked Gallagher and Peter Bahls if they were worried about the future, about climate change and how it would affect all of the restoration work that they have done. Will Gallagher's reference site—the old-growth forest with the ancient trees—and the dream of a valley filled with a wetland and spruce forest be unachievable? Is it even worth having a reference site or a baseline? Was this all in vain? After all, this is a *lot* of hard, often physical, work. Even though the reference stretch of the stream for the project was the old-growth spruce forest of the past, in designing the planting plan in Tarboo Valley, climate change was still taken into consideration. But in the end, Bahls and others working with NWI decided to "plant trees that will grow well now, and . . . work toward a future that doesn't depend on fossil fuels" (Freeman, 2018, p. 48). This approach is a gamble—a hedge for a particular environmental imaginary. But, much like Braverman's (2018) coral scientists, who are restoring reefs against all odds, it is also about hope in the face of potential despair.

Gallagher and I looked out over another section of the valley; this area had young fir trees about twenty feet tall and thick patches of willow along the river banks that had been prepped by NWI's field crew and planted during a Plant-A-Thon, an annual community volunteer work party started by Jude Rubin that has grown into the largest environmental service event in the county. He recalled the labor that went into the young forest we now looked across:

> It was really hard sometimes. I remember at this particular property, we were scalping holes for all of these plants to go in. I scalped *every single one* of these holes *by hand*, by myself. With nobody around except for me (laughs). I was pissed! I was so mad at some points. But, now it's doing good. But if this was just a ditch, all of this moisture would just be pushed downstream. Now it's pretty moist in there. Otherwise, there would be no overwintering habitat, no filtration from wetlands, everything would just go down into the bay. Now we just leave it up to nature, see what happens. But it's now got a good chance, you know.

It is clear to my eye that things have changed a lot in this valley since NWI began their assessment and restoration in 2002. But was climate change on their mind when they were beginning this project? Clearly, it was on Gallagher's:

> Yeah, I thought it was a real threat that people just didn't understand or get. Well, back then, it was "global warming," but I've been thinking and learning about that since I as old enough to think—like fifteen—when people were talking about it in radical circles, because that's where I was at that point. So, I was aware of it, you know. But I also knew that this [restoration] is the conscious effort to fight it. I always thought that, while I was doing it. I mean, you are producing oxygen, and shade, and resilience. Think about our climate now and what it would look like if it was still like that ditched field, and it would just be like a dry tinderbox. Now it's different, it's retaining water, there are plants growing, despite there being a drought. There is moisture in the system. It's healthier. It's a lot healthier. So, we have the ability to change things. For the better. Or, encourage change. And this is just in twenty years. Just imagine it on a larger scale and how well that would do.

Bahls, however, recalled that back in 2002, climate change wasn't at the forefront of his mind:

> Not really . . . it was one of many vague worries floating around, which is not conceivable anymore. But you know, if you are restoring riparian areas and floodplains, you are making the system more resilient to the impacts of things like high stream temperatures and more intense flooding caused by climate change. And if you are conserving forests you are storing carbon. The problem is that these restoration efforts are only a stopgap measure until we get carbon pollution under control. If we don't do something about cutting emissions, we will all be cooked! We won't give a damn what kind of riparian areas we have restored.

I reminded Bahls and Gallagher that this is a relatively large-scale restoration—it's one of the biggest in the region. I wonder aloud to them what it feels like to look at all of this work; at all that's been accomplished here. Their answers echo that of many of the restorationists that I spoke to, and whose voices are heard throughout this book: "It feels awesome," said Gallagher, "It feels really good. It feels like I did something good, you know, that will last a long period of time. Yeah, it feels good. It's a good way to spend a part of a life, you know." Bahls added:

> It feels good, but it also feels frustrating to me. In a sense, while we are remodeling the kitchen, the neighbors are tearing down the entire house for firewood. There's a danger that restoration projects, climate offsets, and other beneficial steps give people a false sense of security that this situation is under control. It's not. If we don't cut emissions drastically, we are rearranging deck chairs on the Titanic as they say. But on the other hand—you do what you can and hope things take a surprising turn and at least some life boats are ready. Doing nothing isn't an option.

Gallagher doesn't begrudge the work he's done, instead, he is driven by climate change and its disastrous effects on salmon to restore even more and work even harder. He believes that "salmon are the reason to restore everything else. They need the habitat and the water. The river is being restored for salmon." While the restorationists working in Tarboo watershed understand that their historic baselines may be anachronistic, they have made a conscious decision not to ignore those old baselines. As I will show in the following chapters, their work, and the work of hundreds of other individual restorationists takes these global dynamics into account through their practices, adapting them so that they can continue to restore habitat and wildlife in unprecedented (climate) times.

4

Emergence

Making Room for the River to Breathe

"**D**ID YOU KNOW that we have been steadily losing snow in the Cascades since at least 1970?" Woodruff asks me, with an urgent seriousness in his voice,

> One researcher says that compared with the annual snowpack of one hundred years ago, the West has lost enough snowmelt water to fill every drop in Lake Mead each year and the loss is accelerating. By 2080 spring snow will be hard to find in the Cascades.

I'm sitting in a coffee shop in Winthrop, in the Methow Valley of north-central Washington. It's a remote, touristy town of a few hundred people. Here, in the off-season between the rush of those seeking world-class Nordic skiing and the deluge of summer hikers, the cafe is bustling with locals who all seem to know one another. They stop in on their way to work for a cup of coffee and a "hello" as they head to their jobs on the ranches and forestlands surrounding the valley. I'm here to talk to a pioneer of beaver reintroduction in the region, Kent Woodruff. He is deeply concerned about the changes that are taking place in this valley, and he is a passionate communicator about climate change, or "extreme events"—the term d'art that he uses in a region with its fair share of climate change skeptics.

For the past ten years, Woodruff has been working to restore riparian streams by reintroducing beavers. The topic makes him come alive. I want to know how he started down this road—and why beavers? He is humble about his work:

> I'm just somebody who saw something on the side of the road and picked it up and started carrying it. As a wildlife biologist, I want to try to improve the landscape. It is a passion for me. It has been a passion in my bones forever.

Even he would admit that the Methow Beaver Project began as a bit of an experiment. Woodruff saw a chance to suggest a change in the way that the US Forest Service was approaching restoration and management of its lands, and he took that chance. As a Forest Service biologist, he was always looking for a way to make a more lasting impact on the landscape; so when the opportunity came along to expand beaver restoration on his district, he just went for it: "When we started, we weren't sure if we could make a difference at all," he says. That was back in 2008, and since then, the Methow Beaver Project has grown into a nationally renowned example of the successful restoration of ecological processes through beaver reintroduction. Project managers and volunteers have relocated hundreds of "nuisance" beavers, from lowland areas where they are viewed as destructive to farmland, to headwater areas on public lands managed by the Forest Service. There, the beavers can get to work doing what they do best—complexifying streams and reconnecting them to their floodplains. In short: making a mess. Up in the headwaters on the Forest Service land, this isn't as much of a problem as it is down in the valley, where private property owners might not be excited about this kind of ecological "mess."

As Woodruff tells me about the origins of the project, we are interrupted from time to time as he greets a familiar face walking into the coffee shop. In the background, I overhear conversations with the barista. "What do you think of the river?" "Highest I've ever seen it." "I think it's running at 7,500 CFS."

The River. You can see it from the back porch of the coffee shop. Whole trees with their roots and soil still attached are being flung down its course. The Methow River is the talk of the town today, as it has been for the week I have been staying here. As I drove up the Columbia River from Wenatchee on my way to the Methow Valley, I saw a vivid change in the Columbia as the crystal clear, blue-green water of the main stem began to mix with a chocolate froth. Plumes of milky-brown water and forest debris roared down the river to its confluence with the Columbia. Logs, sticks, and everything in between were pulled downward by the heavy flow. I had never seen the Methow like this, in all its meltwater glory. The tumult and debris looked dramatic to my outsider's eye. The precipitously rising waters appeared to be a disaster unfolding. But as I spoke to the locals, I learned that, while this was considered "high" water and was close to flood stage, it was also simply the way the river works. Every few years, high waters like these race down from the mountains as the snow melts. Sure, this might tear out a few culverts and cut some deeper turns, but this is what rivers like the Methow, which are driven by snowmelt, are "supposed" to do.

Nevertheless, the River was the talk of the coffee shop. People were excited to see it so high and dynamic as it raced by the back door, carrying whole trees with it. But as dramatic as it looked, it wasn't something to worry about. Woodruff picked up on the banter of the coffee shop:

> The river that we are sitting next to should have the opportunity to move all over the valley. As we sit here, the river is coming up and it has been doing some exciting things in the last couple of days. We are fighting it as much as we can and trying to keep it confined, but salmon want that river to breathe.

To dramatize, Woodruff takes a moment to breathe in and out slowly—he is gifted with a dramatic flair. "And," he continues after a moment, "room to breathe means that it needs the entire floodplain—not riprap to constrain it. . . . Our big hydrograph here is a key feature of the ecosystem function." As process-based restorationists would attest, embracing the dynamism of nature is something that ecological restorationists have been advocating for decades. In anticipation of climate change, these practices are expanding to encompass larger scales and more interdisciplinary perspectives. The epistemic community of restorationists openly fosters a culture of experimentation where "unruly complexity" (Taylor, 2005) is encouraged. This is leading to the emergence of new ideas and constitutes an adaptation for dealing with environmental change and uncertainty.

Back in 2008, Woodruff's "try anything" attitude helped launch the expanded Methow Beaver project. Since then, the project has brought a diverse set of people in the Methow community together in an effort to restore watershed processes by using the work of beavers as a restoration tool. Even as we finished our coffee next to the surging tumult of Methow River snowmelt, Woodruff spoke about how the uncertainties of climate change have affected his own work: "When you're on this roller coaster you want to be like, 'Whoa! Slow down.' It's very difficult. It's very difficult to try to get some scientific answers in such volatile systems." In this chapter, I explore one of the strategies that restorationists are using to deal with this uncertain environment—the "beaver dam analog." Strategies like this occur at both the individual and the cultural level of scientific practice; and these changes are influencing the broader culture of ecological restoration in the Columbia River Basin by breaking down divides between engineering-based and process-based restoration to lead to new ideas fostering adaptation to the uncertainty that climate change is bringing to the field.

EXPERIMENTATION AND EMERGENCE

There are people like Woodruff throughout the Columbia River Basin: people who are experimenting with innovative ways to restore the ecosystem processes that were (sometimes completely) lost as the river and its headwaters were developed and confined by roads, irrigation, and hydropower projects. Some restorationists are even working to engineer ecosystem processes into places they never existed. As described in previous chapters, process-based and engineering-based restoration have been in tension throughout the evolution of the field, and many people still firmly situate themselves in one "camp" or another. One reason for this divide is that many restorationists view the legacy of engineering "solutions" as a major contributor to the problems that got rivers into trouble in the first place. Engineers constrained and straightened river channels, coinciding with habitat loss. Woodruff, for instance, juxtaposes his desire to "give the river room to breathe" with the riprap and concrete that constrains it.

Yet ecological restoration, especially of rivers, owes much of its roots to hydraulic engineering—and designers and engineers still play a major role in restoration planning, science, and implementation in the Columbia River Basin. There are also those who think that, while process-based restoration might be the ultimate solution and beavers might be a step in the right direction, restoring ecological processes takes a long time; too long, in fact. They worry that efforts to restore process may not be enough or may not happen fast enough to mitigate the effects of climate change that are already taking place. After witnessing the devastating salmon die-offs of 2015 in the basin, many people are willing to implement innovative measures—whether engineering *or* process-based—so that this type of ecological disaster is less likely to happen in the future.

In practice, many restorationists find themselves pulling from both toolboxes—process-based restoration tools and more technologically-based engineering solutions to try to "hedge" the uncertainty that climate change introduces. These myriad strategies for dealing with uncertainty and adapting to climate change are being fostered within epistemic cultures that embrace experimentation; they gamble that if enough restoration tactics are tried, something will work. As climate change takes hold, the collective goal is increasingly becoming to bring fish back by creating habitat *by any means necessary*. While some restorationists still disparage those in "opposite" camps, the divide between the two cultures is breaking down, as approaches once seen as incompatible come together in practice when restoration projects are implemented. One example of this is found in the beaver dam analog—an engineered dam structure that by mimicking a beaver dam, often entices a beaver to take it up as its own.

Beaver dam analogs, or BDAs as they are called by practitioners, are "channel-spanning structures that mimic or reinforce natural beaver dams" (Pollock, Weber, & Lewallen, 2015, p. 82). At their most basic, they are human-constructed beaver dams, often built using pneumatically-pounded posts and then weaving branches of cottonwood or willow behind them. After adding sediment and other materials, BDAs have been shown to increase ecological function in much the same way as natural beaver-maintained lodges do in a landscape (Pollock et al., 2015b). The key to their success is to locate a suitable location by "understanding the hydrogeomorphologic context" (p. 82), making proper planning and design essential (Pollock et al., 2015b). The amount of ongoing human intervention needed to keep BDAs intact varies: these "starter dams" may entice beavers to an area, thus beginning a process-based restoration, or they may require ongoing maintenance if natural processes do not take over (Pollock, Weber, & Lewallen, 2015). BDAs are increasingly common as a tool for restoration in parts of the basin that could benefit from the initiation of beaver-driven process-based restoration.

I argue that the two philosophies of restoration: process-based and engineering-based, come together and find common ground through BDAs. While these constructs are usually fairly simple structures made to mimic a beaver dam, they are increasingly drafted into engineered restoration designs. Much like beaver dams, these small-scale alterations on a waterway can have large-scale and long-term effects. This embrace of both natural processes and engineering coincides with the rise of the Stage 0 restoration strategy—described fully in chapter 2—in restoration. While BDAs have some resemblance with the engineered in-stream structures that have been common in restoration, many engineering firms are embracing and adopting BDAs into their designs instead. In this way, they are both familiar and unfamiliar; they release control from an engineered structure—often held in place by steel cables or concrete—giving agency and control to the beavers and to the ecosystem instead. This demonstrates how even highly entrenched epistemic cultures can shift, as unfamiliar concepts and methods are tolerated in order to see what happens in times of uncertainty such as these.

THE WORK OF BEAVERS

By employing the tools of paleoecology, researchers have found that beavers and salmon evolved together (Davies & Gibling, 2011). The oldest record of beavers hails from about 30 million years ago, and the oldest beaver dam from around three million to five million years ago (Davies & Gibling, 2011). Some restorationists believe that restoring beavers is a relatively fast way to restore fundamental ecological processes—and eventually Stage 0—to large areas. Beaver restoration advocates like Woodruff conceptualize restoration on an evolutionary time scale. To them, restoration is about restoring the processes that existed before the river was transformed by development—and that process is going to take a long time. As one beaver restorationist put it, "we need to do more than throw some sticks in a crick." He is referring to the common restoration tactic of engineering log jams in a river from large woody debris. He'd like to see beavers doing this work instead.

The idea behind using beavers to create wildlife habitat is not new, and efforts to restore beavers to ecosystems have been taking place since at least the 1930s (Wheaton et al., 2019). These efforts, some of which are outlined in Heter's (1950) now legendary article: "Transplanting Beavers by Airplane and Parachute," tells how managers employed a novel way to deliver beavers to the headwaters of the Upper Snake River for the purposes of creating ponds for waterfowl. Unfortunately for today's beaver restoration scientists, these projects were not monitored, and so it is unclear if they were successful. Yet even before those early efforts: the Columbia River Basin has been continuously developed by Euro-American settlers since the 1800s, and it has been humans—not beavers—that have drastically altered the hydrologic regime. As settlers moved in, many wetlands were drained, and beavers were either trapped for their pelts or killed because they were seen as a nuisance causing flooding and property destruction. Through hunting and trapping, they were nearly extirpated from the Pacific Northwest by 1900 (Bouwes et al., 2016). The removal of beavers from Pacific Northwest ecosystems also succeeded in reducing the places where salmon can spawn, rear, and grow. This is because beavers are essential ecological engineers, creating diverse habitat through their life histories and behaviors.

Beaver reintroduction is process-based restoration at its most basic: add beavers, and they start the process of floodplain reconnection by creating biogenic dams. "Biogenic dams" are dams created by living organisms—beavers, dead and living trees, roots, and vegetation alter the movement of sediment and water, forming networks of pools and channels that are ideal salmon rearing habitat (Pollock et al., 2014). Beavers not only create fish habitat: in the same study, Pollock et al. (2014) also monitored the effects of beaver restoration on groundwater and found that water over a kilometer away was affected, with bigger flows and colder upwelling occurring downstream. Other studies have demonstrated increased steelhead density, survival, and production after beaver dam analogues were installed (Bouwes et al., 2016). Beaver restoration is founded on the idea that creating these biogenic dams will reconnect floodplains and produce salmon-rearing habitat. This is habitat that is diverse and, most importantly, connected to the temperature-mediating abilities of groundwater. In the context of climate change, this is critical: especially in areas like the Methow Valley, where water temperatures and flows are a limiting factor in salmon survival. The additional water storage that beaver dams provide in headwater areas could therefore potentially help compensate for the loss of water storage in the form of high-elevation snow due to climate change (Lawler, 2009). Some restorationists believe that beaver could help create this storage—something that is now provided through reservoirs, the very structures which harm salmon migration.

Jeff Oveson has also experimented with beaver restoration and recently retired as executive director of the Grande Ronde Model Watershed. He explained:

> The obvious conclusion is going to be that we need to find a way to store water. The motive for some people is that we need to justify building a reservoir. People still don't seem to understand that reservoirs are horribly expensive and problematic.... It's expensive. Beavers are really good at storing water. They maintain the storage. There are no pumps. That is probably an oversimplification but... beavers move a lot. A lot more than people think.

By incorporating BDAs into restoration designs, process-based restoration is becoming recognized as an important avenue to counteract the effects of climate change. On paper, many restorationists don't explicitly incorporate climate change data or mitigation measures into their designs: but when asked how they are addressing climate change, they often cite process-based restoration as their strategy. For example, Jennifer Molesworth, a fisheries biologist has also been working in the Methow for almost two decades, is a sub-basin coordinator for the Bureau of Reclamation, and a straight talker when it comes to her experience in the sub-basin. She explained,

> Our approach is more based on physical process. While there is temperature data that gets considered, and we are always looking for groundwater sources and refugia—this is in part a climate adaptation—we also just want to restore natural river processes and connections. I think our underlying philosophy tends to be: where we can restore natural river function and reconnect floodplains and groundwater sources, that this will accomplish the goals of preparing for a changing climate. And, if we put the river into its naturally functioning state, it will be able to maintain itself and we won't be in a maintenance mode all the time taking care of artificial things. It will be in a state that, theoretically, fish are adapted to.

Beavers are capable of restoring long stretches of stream—something that is costly to do by hand and often impossible to do in remote headwater areas. By reintroducing beavers to areas where they no longer live, advocates of beaver reintroduction programs are hoping to glean these beaver benefits. The Methow Beaver Project relocated 274 "nuisance" beavers between 2008 and 2015. They were brought to 61 sites and they established themselves at 30 of these (Woodruff, 2016). This high rate of success could be due to the care that project participants and volunteers took in the relocation process—pre-building partial "artificial lodges," providing forage, and making sure that beavers pairs are compatible. Relocation in other parts of the basin has also been successful, with varying rates of success.

Despite its historical roots, beaver restoration for the purposes of salmon recovery is fairly new and restorationists are only beginning to monitor the results and perfect the methods. The engineering paradigm for restoration remains dominant in many regions and organizations, and skepticism about beaver restoration also exists. While many people are beginning to incorporate process-based techniques into their restoration designs and planning, the legacy of engineering-based restoration is still strong. Molesworth has seen this skepticism pointed at process-based restoration:

> Initially, I think, there was a lot of skepticism over restoration work and why we were doing it because people have seen the government agencies tear out wood and straighten streams and riprap streams, and they kind of thought, "Oh. That is how you handle a stream." So, we have this big, huge paradigm shift that we are working through to get people's eyes and brains retrained on what a healthy functioning river actually looks like. . . . Right now I think we are still in the phase of: "You want to put wood back in!? You want to take out a levee!?"

Shifting a culture from one based on engineering to one based on process is particularly difficult in river restoration, where floodwaters and erosion can put infrastructure, and even lives, at risk. Molesworth pointed out the practicalities: "There is a reality of working where everyone is. You have irrigation systems and houses and so we have to keep back from that." Her job brings her into contact with stakeholders who see the river in many different ways, and some view it as a potential threat whose floodwaters are dangerous to life, infrastructure, and property, even if they do want to see fish return.

There is another school of thought that engineering-based restoration is necessary to "bridge the gap" between conditions now and a climate-changed future. As Molesworth pointed out, oftentimes it comes down to the wire, and people say:

> Let's build a fairly contrived or artificial habitat that doesn't necessarily address process. You are just trying to make more fish. And you are trying to make more fish immediately so that you can bridge the gap so that they don't go extinct and you still have fish around later, and then worry about the other stuff [process] on another separate timeline. My preference is for restoration of natural process; but if we don't have the time for that and species are going to go extinct and we have to do something: that is a really tough choice right? Culturally and societally.

Putting beavers—an animal that is still often viewed as a pest—back into the environment doesn't come without controversy within the restoration community; but as more studies that demonstrate its efficacy become available, the culture is beginning to shift. Paradigms don't necessarily shift quickly. Nonetheless, experimenting alongside "nature's engineers" is facilitating the emergence of new ideas and potential solutions that can be added to a climate adaptation "toolbox" for restorationists.

EMBRACING UNRULY COMPLEXITY, BECOMING INTERDISCIPLINARY

Due to the success of these projects, beaver restorationists at the US Forest Service and other agencies have been hosting workshops in order to train ecological restorationists in the science and art of beaver reintroduction throughout the Pacific Northwest. The workshops spread information about how to restore using beavers, but they also help practitioners to hone these new practices by gathering feedback on local successes and difficulties. *The Beaver Restoration Guidebook* (Pollock et al., 2015b) breaks the process of beaver restoration down into simple steps:

1. Identify suitable habitat (often using remote sensing).
2. Assess current beaver population status and distribution.
3. Evaluate individual release locations.
4. Pursue acquisition of beavers.
5. Collect information about beavers captured (or recaptured).
6. Care for beavers temporarily and ensure that beavers are grouped as families—or compatible units with both males and females.
7. Prioritize and prepare release locations.
8. Deliver beavers to selected sites.
9. Conduct follow-up monitoring and provide support.

While this step-by-step process seems fairly straightforward, beavers themselves often behave in unpredictable ways. Scholar Cleo Woelfle-Erskine (2017) calls beavers, "stochastic transgressors against Manifest Destiny engineering projects" (p. 5), highlighting their transformative abilities—and the way they alter landscapes differently than engineers—as they become collaborators with restorationists in the transformation of rivers (Woelfle-Erskine, 2017). Interestingly, the "messiness" and unpredictability of beaver reintroduction is not a hindrance to restoration but instead a useful strategy for adaptation. It introduces a place where the emergence of new tactics to deal with climate change are being fostered.

According to some restorationists, too many restoration plans are based on ideas about how streams *should* behave: for example, with a two-year flood event, a five-year event, and so on. Some models still also assume a steady state, or unchanging balance, in an ecosystem. The engineering-based approach is still viewed by many as the dominant approach, and, as dominant, something that needs to be upset through process-based and low-tech restoration (Wheaton et al., 2019). Engineering-based techniques may include channel reconfiguration and engineered log jams or other in-stream structures; but many see process-based approaches as a complement to these more intensively constructed approaches. Wheaton et al. (2019), for instance, believe that the "precisionism"—or the need for certainty and stability—and the high cost of engineering-based restoration are limiting the scale and scope of many projects. Instead, they suggest complementing engineering-based approaches with low-tech, process-based tools such as BDAs to "kick-start" ecological processes (Wheaton et al., 2019).

The problem is that biogenic dams, either BDAs or the ones that beavers construct, are *meant* to cause instability and act unpredictably, exactly the opposite of what an engineered system is supposed to do (Wheaton et al., 2019). It is also, frankly, difficult to model and predict beaver behavior. Woodruff explained that monitoring the effects of beavers can be daunting in part because beavers can cover a lot of ground:

> The variables involved the challenge of conducting a scientific investigation where your treatment is: "let's put some beavers here . . . and let's put some beavers here . . . and let's put some beavers here . . . and then . . . oh wait where did they go? We just put them right there where did they go?!". . . They often move where we did not expect. We also have landscape impacts and mortalities that we didn't expect.

Despite these difficulties, beaver restorationists have measured positive fish response to the restored wetlands and complex pool systems that beavers create, including increasing survival of juvenile steelhead thanks to increases in both the complexity and quantity of their habitat (Bouwes et al., 2016). Although results such as these are beginning to be published, as one restorationist pointed out, you aren't necessarily going to be able to see the results of this ecosystem recovery "in a publishable time frame." The long timescale of ecological recovery makes it difficult to measure success in beaver restoration; still, short-term changes have been demonstrated.

Yet, even if it is difficult to measure and predict the results from a beaver reintroduction, or a BDA placement, unpredictability and dynamism is exactly what process-based restorationists want to reintroduce to the river. This means that tolerating some imprecision and unpredictability is necessary. Beaver restorationists are doing this work because, just by "trying things out," new ways to address the effects of climate change might emerge. Woodruff agrees:

> When we started we weren't sure if we could make a difference at all. We weren't sure if it would have an influence—and it didn't start out as a climate change project. But then I got to realize that we are struggling [with lack of water] and at least some places below beaver dams continued to run when the rest of the stream went dry. Then it became obvious that this is something that could allow streams to keep flowing as we continue to lose snowpack.

This is an example of the ways in which restorationists have come to embrace unruly complexity, and the emergence of novel concepts that occurs with experimentation as they hedge their bets against climate change. Despite a growing number of monitoring efforts and resulting publications that show efficacy of methods like BDAs, there is little evidence to "prove" that introducing beavers will mitigate climate effects. Even without this evidence, the epistemic culture of restorationists is, for the most part, beginning to consider beaver restoration as a way to confront environmental change.

SHIFTING EPISTEMIC CULTURES: EMERGENCE AND ADAPTATION

Theories of scientific practice and social and cultural change often employ a concept of "emergence." The term is used to describe novel concepts or properties that arise out of the interaction between different, more fundamental, concepts or properties. The term bridges disciplines and has been applied to dynamics at multiple scales of science, from biology and physics to social relations and the production of knowledge. Emergence is one of the key dynamics in conceptual models of socio-ecological systems and in adaptive management (Walters & Holling, 1990). Emergence of new practices, ideas, and situations is driven by the interaction of parts in a complex system.

Both empirical and theoretical work on socio-ecological systems has illustrated the dynamics that give rise to emergence in these systems. For example, Ostrom (1990) documented the emergence that occurs through self-organizing governance systems and the ways in which these dynamics can foster innovation in natural resource governance. Social theorists also consider culture to be the "surface of emergence" from which humans structure their discourse and actions in epistemic work (Foucault, 1972). Scientific practices emerge in a dialectic exchange: as scientists work, they come up against "resistances," whereby they must modify their hypothesis, their ideas, or their practice itself in order to carry on (Pickering, 1995). These modifications have been called "accommodations," and they lead to the emergence, or adaptation of scientific practice (Pickering, 1995). This dialectic process is conceptually similar to emergence in the adaptive cycle of panarchy: where cross-scale interactions between parts of a system lead to emergence of new practices and cultures (Gunderson and Holling, 2002). By understanding epistemic culture in this way, we can see how different cultures will lead to the emergence of different concepts and ideas. For example, as we have seen in previous chapters, an epistemic culture that is more technoscientific is also more likely to develop engineering solutions to address environmental problems, such as hatcheries for fish passage in the case of Columbia River salmon restoration.

Although process-based and engineering-based restoration cultures may exist in contradiction, they also exist in tandem. This is a common feature of science in general, although it is often overlooked in an aim to synthesize science into distinct epistemic cultures. And, although Kuhn (1962) argued that scientific cultures form through unitary conceptual frameworks in the form of paradigms, others have argued (Hacking, 1992) that scientific work is actually conducted through heterogeneous, patchy cultures and practices that are not uniform conceptually, but instead "mutually" adjust to each other as work is carried out and environmental and social conditions change.

This is the case in ecological restoration, as demonstrated in the cultures that have formed around process-based and engineering-based ecological restoration. As ecological restoration has matured as a field, many of the lines that were "drawn in the sand" seem to be gaining less traction as the new common foe of climate change enters the picture. This can be seen in the way that process-based and engineering-based restoration are coming together through the concept of BDAs. Janine Castro, a long-time restorationist and educator, who has helped develop pioneering beaver restoration educational programs throughout the basin, described the shift this way:

> I think the big change is that we are going from really focusing on channel structure and not paying attention, to things like macroinvertebrates or primary productivity. In our program, we have the physical processes class and the ecological processes class, and what we are seeing is that they are coming closer and closer together. Especially when it comes to beaver. They create dams, but they are organisms, so they are engineering organisms. There's this really interesting kind of connection between these disciplines and they are coming much closer together and recognizing that the separation is pretty artificial. I think people are becoming interdisciplinary.

These interdisciplinary strategies are not only a way to deal with the mess of natural systems, or what has aptly been called unruly complexity (Taylor, 2005). Unruly complexity describes the ways in which ecological processes are not easily generalized and bounded into discrete systems; instead, they are replete with intersecting processes that are difficult to predict (Taylor, 2005). Ways of working that tolerate complexity foster cross-scale and interdisciplinary interactions so that new concepts and ideas are more likely to emerge. These very properties also drive change in scientific culture (Pickering, 1995). Yet strategies that embrace experimentation and unruly complexity must be supported through epistemic cultures that value these kinds of epistemic virtues, fostering them through scientific practices and institutions that allow the flexibility to make mistakes. A willingness to experiment with new restoration strategies, such as BDAs, that cross the process/engineering divide demonstrates that epistemic virtues and cultural norms that tolerate these risks are more favorable to the alternative risks of not trying anything at all, the do-nothing approach that lets climate change drive failure. These shifts point to an evolving diversity of restoration efforts, including improvisation and experimentation, that allow restorationists to foster the emergence of novel solutions and deal with the uncertainty and immediacy of climate change's influence and effects.

CONCLUSION

In dealing with uncertainty and change in their work and in the field, restorationists are embracing a culture of improvisation and experimentation in which they engage with unruly complexity, as well as diverse disciplinary concepts. Because multiple cultures will contain multiple surfaces of emergence, an epistemic community that embraces multiple disciplinary cultures is more likely to contain more concepts and tactics, and thereby more avenues for solutions to potentially emerge. Dealing with the uncertainty of climate change by fostering emergence is therefore becoming an adaptation for dealing with climate change in restoration. These strategies can be seen in beaver restoration, as restorationists employ novel restoration design strategies and create new goals. Through the work of beavers, ecosystem processes are set in motion, unpredictability is encouraged, and emergence is thereby fostered through embracing the dynamism of nature.

In this chapter, I have outlined the ways in which fostering emergence is an adaptation that restorationists are using to deal with climate change. Using techniques from both engineering- and process-based restoration, multiple cultures are engaged, creating more opportunities for new concepts and practices to emerge. One of these practices is beaver restoration. Through combining the work of beavers with engineering-based approaches, the novel concept of the beaver dam analog emerged.

Yet, fostering emergence is also a way to embrace unpredictability, which is what river restoration is, in many ways, about—giving the river "room to breathe." The concept of emergence in scientific culture and practice helps remind us about the particular qualities of human and nonhuman agency that restoration entails. Restoration is about "doing things" to nature, and then seeing what nature does back. Therefore, thinking about the emergent qualities of ecosystems and human agency is useful in understanding what drives change and adaptation. Interactions between components and scales are a constant dynamic within an ecological system, and a hydrologic system, such as the Columbia River Basin, can be highly variable, or stochastic. In other words, the river itself is constantly "doing things." No matter how much some humans would like to (and try to) control the actions of the river, they cannot entirely do so. Therefore, the material world has a certain degree of agency that acts on scientists and alters their practice. Appropriately to the problem of climate change, Pickering (1995) uses the example of "the weather," to illustrate the agency of nature, in which "winds, storms, droughts, floods, heat and cold . . . engage with our bodies as well as our minds" (p. 6). Scientific practices, according to him, are a "continuation and extension of coping with this material agency" (pp. 6–7).

This "rebalancing" of agency is particularly helpful for conceptualizing how restorationists deal with climate change in their practices. As scientists and the environment interact through time, scientific practice takes on an emergent quality. As new problems arise, new solutions emerge. Just as in adaptive management of socio-ecological systems, uncertainty always exists; and this uncertainty is accepted as a dynamic to be fostered, not contained. In addition to the actions of beavers, the river and its ecological processes, the salmon, the ocean, and climate change, all play a role in this emergence through temporal and spatial cross-scale interactions. This playfulness tests and stretches what restoration design can accomplish. It is not only found in beaver reintroduction, but also in innovative engineering solutions such as BDAs, that may lead to unexpected outcomes, allowing new ideas to emerge. As one regional restoration coordinator explained in reference to a project with some unintended outcomes:

> Nature will always expose the hidden flaws. That is the nature of nature. So, if there is a flaw in an approach sooner or later it will become obvious . . . Good scientists laid [the projects] out and they had a good plan, but they just couldn't execute it. Too many things kept popping up and interfering in fundamental ways. So maybe the lesson is the chaotic unpredictability of climate change.

Daston and Galison (2007) compare the emergence of new epistemic virtues to an avalanche: "at first, a few tumbling rocks, falling branches, and minor snow slides amount to nothing much, but then, when conditions are ripe, individual events, even small ones, can trigger a massive, downward rush" (p. 49). In a similar way, the epistemic virtues that are emerging to cope with climate change are only just beginning to surface, but they are nevertheless present in the way that restorationists in the Columbia River Basin talk about their work.

Different scientific virtues and cultures coexist with one another, and as time passes, more of them accumulate and add to the mix of possibilities (Daston & Galison, 2007). Individual scientists and managers draw from these virtues and cultures in their work; and for restorationists, this adaptation that allows multiple cultures to coexist is becoming a way to deal with the unexpected: with climate change.

To be sure, some restorationists are loyal to one camp or the other and see the split between process-based and engineering-based restoration as embodying important differences in assumptions and even ideologies. Yet at the same time, as I described through the story of the development of ecological restoration in the Columbia River Basin in the previous chapters, these virtues and cultures cannot be attributed to individuals alone. Instead, they are developed through a "collective empiricism" that has evolved over time (Daston & Galison, 2007). When restorationists restore natural processes by reintroducing beaver to a watershed, they are drawing from a tradition of process-based restorationists that have worked toward similar goals for decades. Similarly, when restorationists draw on engineering principles to engage in habitat restoration, they too are drawing on a history of environmental engineering that has developed over time. When individuals draw from these scientific resources, they are building on collective histories of conservation work in the basin, each with particular epistemic cultures and virtues that align with them. The two modes of conservation work represent different ways of seeing nature, different ways of knowing, and different tactics for dealing with change. Yet as climate change becomes more tangible, people are beginning to work towards what they describe as a "common goal"—restoring salmon to the basin by any means necessary.

Beavers, often called "nature's engineers," break down the engineering/process divide as they engineer their own dams throughout the headwaters. In the process, "success" is also redefined, as learning something—anything—becomes more important than achieving a particular outcome. As Janine Castro pointed out,

> You know, it's not black and white. We don't have "success" and "failure." We are often working in the gray areas. One of the things that I am really pushing is this idea of success where I'm not interested in success, which indicates an endpoint or target, but that, really, it creates oscillations of the system and it has a heartbeat. If you don't have that, and it never changes, then it's never going to be okay.

Enrolling beavers in restoration embraces the imprecision, unruly complexity, and dynamism that ecological restoration involves: an emergent moment of epistemic cultural change that is helping restorationists adapt to a changing climate.

5

Acclimation

Rethinking Monitoring for the Future

RESTORATIONISTS IN THE Columbia River Basin are not only tasked with restoring ecological function and processes to an altered—and some say "broken"—river; they also need to determine whether or not their restoration actions are working. In the early 1990s, when restorationist Anthony Bradshaw (1987) framed restoration as the "acid test" for ecology, he emphasized the extent to which scientific experimentation is embedded within the more applied, practical matters of restoration design and implementation. However, as described in the previous chapters: in the Columbia River Basin, salmon habitat restoration is not only a scientific endeavor, it is also driven by the Endangered Species Act's mandate to recover endangered species. Because of this relationship, change in scientific practice often occurs not necessarily as an adjustment to a new knowledge or technology but in response to a change in policy. The ESA requires those active in recovery efforts to demonstrate the efficacy of their efforts through specific metrics and monitoring practices that together determine much of the monitoring work itself in the basin. At the same time, monitoring practices must also adapt to environmental changes, which can complicate restorationists' ability to design monitoring protocols that will continue to provide the information needed to move forward as conditions change along with the climate.

This dynamic between science and policy is not unique to the Columbia River Basin but is found around the globe in environmental regulatory regimes that call on science—and climate change is confounding this dynamic globally. While monitoring constitutes a major scientific effort in the Columbia River Basin, monitoring programs have been some of the most problematic and often neglected aspects of restoration (Bash & Ryan, 2002). Many local and regional restoration projects conduct their own monitoring and several major, large-scale monitoring programs have also been organized by NOAA Fisheries, the Bonneville Power Administration, and the Northwest Power and Conservation Council. These monitoring programs are often "fish-centric," meaning they are counting "fish in and fish out" of the rivers. Individual fish are either tracked using PIT (passive integrated transponder) tag arrays, or counted in "snorkel surveys" or by electrofishing. Other monitoring programs focus on habitat, water quality, and potential productivity, measuring changes to in-stream temperature, stream flow, channel morphology, or percentage of vegetation shade cover and large woody debris.

What to measure, how to measure it, and especially how to tell if restoration is making a difference in habitat conditions are all topics of keen debate in the restoration community. Restorationists throughout the basin hold strong opinions about monitoring, and oftentimes these opinions result in contentious debates at conferences and meetings. Some believe that monitoring efforts have been lacking, while others think that the programs are focusing on measuring the wrong things. Still, others think that monitoring is a waste of money entirely. These conflicts and tensions often center around anxieties about how to deal with the large-scale and ambitious goal of the recovery effort: restoring salmon populations to the Columbia River. These concerns are now compounded by the uncertainty inherent in drawing baselines and conducting monitoring in the context of climate change—chapter 3 described this challenge.

In this chapter, I explore how, despite these challenging circumstances, restorationists are developing robust monitoring strategies to inform restoration design and measure restoration success in a climate-changed environment. To explore this set of strategies, I use the word "acclimation" to describe a process of adjustment in which restorationists are forced to use a pragmatic approach to understanding and knowing the river. I describe how restorationists do this through acclimating their scientific practices by changing the ways that data are collected, including learning to use "trained judgment" to recognize and understand changes in the environment. Using the example of a regional monitoring program, I argue that restorationists are being forced to acclimate the knowledge infrastructures that scientists rely on, and that they are building today, as they adjust their metrics, standards, and classification systems to adapt to climate change. These adaptive knowledge infrastructures are the key to enabling scientific work—including environmental monitoring—to continue to be effective through a changing climate.

COMING TO "KNOW" THE RIVER

When salmon habitat restoration monitoring began at a programmatic level in the Columbia River Basin, it mainly focused on measuring the physical changes to stream morphology caused by instream restoration. Only later did these monitoring programs evolve to evaluate restoration of ecological processes. Most early monitoring studies were also aimed at understanding individual sites and were not meant to be widely applied (Roni, 2005). However, as adaptive management expanded into a basin-wide recovery effort, monitoring became even more important. Restorationists needed to be able to evaluate actions and to make adjustments to their management efforts if needed (Roni, 2005). While large-scale monitoring has been from the beginning a necessity in the Columbia River Basin, it has taken decades to create a comprehensive monitoring program. Even now, the practice of monitoring for ecological responses to habitat restoration lags far behind the monitoring programs established to measure individual metrics such as number of fish, miles of restored stream, or number of instream structures. One problem with these metrics is that they are often focused only on short-term benefits. Meanwhile, since the 1990s, restorationists have advocated for a more comprehensive, long-term monitoring program that takes a holistic approach to understanding underlying ecological changes (Beechie & Bolton, 1999; Naiman, Bilby, & Kantor, 1998; Roni, 2005).

Although the purpose of monitoring habitat restoration is to understand how restoration actions affect the environment—and by extension, how those actions impact fish—restorationists in the Columbia River Basin refer to several different monitoring "goals" or approaches. *Implementation* monitoring focuses on how much restoration is being done "on the ground." *Effectiveness* monitoring seeks to understand whether or not restoration is working to alter the environment in desirable ways. *Baseline* monitoring is focused only on characterizing the existing conditions to enable future comparison. *Status* monitoring looks at the variability of conditions across an area; and *trends* monitoring looks at changes through time, taking a "pulse" of the environment in order to provide a baseline and quantify changes taking place. Finally, *validation* monitoring evaluates whether a hypothesis about a restoration's effects was correct (Roni, 2005). Each of these monitoring approaches has met with different success rates as the recovery effort has advanced, and—in the end—a combination of monitoring types is often used.

MONITORING PROGRAMS IN THE COLUMBIA RIVER BASIN

The main source of funding for habitat monitoring in the basin is provided by the BPA in order to understand whether or not their mitigation efforts are working (Bonneville Power Administration, 2019). To meet this need, funding has been directed toward three main monitoring programs: the Integrated Status and Effectiveness Monitoring Program (ISEMP), the Columbia Habitat Monitoring Program (CHaMP), and the Action Effectiveness Monitoring of Tributary Habitat Improvement (AEM). Each of these programs has different purposes, but they all work together to some degree. For example, ISEMP develops monitoring protocols so that standardized data collection methods can be used across the Columbia River Basin to determine the effectiveness of habitat restoration actions within CHaMP study areas. An additional, related program has worked to establish intensively monitored watersheds, or IMWs. The IMW program is an attempt to understand the effectiveness of restoration at the watershed, or population scale, instead of the more commonly targeted and smaller "reach" scale: but it has been challenging to implement due to lack of consistent funding and coordination (Bennett et al., 2016).

The Challenges of Monitoring

The words of one frustrated informant, a researcher who works on comprehensive restoration planning, summed up many of the problems surrounding monitoring:

> The Northwest Power and Conservation Council has spent $66 million in the past decade on monitoring, and they are about to abandon virtually every one of those programs. So, how does one cost-effectively monitor the stuff that fifty years from now people will say, "Yeah you made the right choices about what to monitor, and you made those choices so you could actually monitor them for fifty years?" . . . If our society had the will to spend a billion dollars monitoring all of the Columbia Basin for all of those qualities for the next five decades, sign me up. I'd be willing to pay my taxes to support that, but it's not going to happen.

How does one know what to monitor? Indeed, how does anyone know what will remain a useful baseline to have in fifty years? Are the same things that matter to society today going to matter in the future? What if it is all too little too late, and monitoring is a waste of resources? And after all that: who is going to pay for it all? The person quoted above is not the only one who is concerned: this kind of adaptive thinking about what kinds of data should be collected now in order to be useful in the future is not just an anxious worry of a frugal scientist. These questions need to be considered seriously in order to create an adaptive ecological restoration science that is enduring, useful, and cost-effective into the future.

For example, the complexity of overlapping cycles of fish migration, ocean cycles, and global climate make monitoring a multilayered and multiscale endeavor. Further, these cycles are often at odds with the cycles and timescales of politics, funding, and especially, careers. Restorationists are keenly aware of this mismatch. Reminiscing about a three-decade career of implementing and monitoring projects, one government agency restoration ecologist told me:

> Even if you have a perfect method of monitoring all of those things, the interaction of them is complicated. You might be doing great habitat work, which theoretically will produce all sorts of benefits and abundance; but the adults [salmon] are going back out into a terrible ocean in an eight- or ten-year cycle, which you have to wait through to see the benefit. And humans have a really hard time being patient. You might only get two or three of those cycles in your whole scientific career. How do you make the managers wait through that to see if it really paid off or not? It is hard.

Restorationists have had to adapt to these complexities in scale in innovative, and both simple and complex, ways. Importantly, they have had to do so quickly, accepting uncertainty as a given.

Short-term changes can be difficult if not impossible to detect when the life cycle of a salmon is five to eight years and much of their lifetime is spent in the ocean, thousands of miles away from a restoration project site. This means that there are a multitude of factors affecting their survival, not just the quality of habitat in the rivers. Because of this complexity and the long timescale, restorationists especially struggle with effectiveness monitoring (Paulsen &Fisher, 2005). After completing a major restoration project in the headwaters, about three hundred river-miles from the ocean, one restoration manager pointed out that the project,

> could make all the difference in freshwater survival but . . .
> if we have so many Caspian terns and double-crested cormorants and so many harbor seals and a "blob" out in the ocean, who cares if you have a five-fold increase in your population, if your population is nothing.

It is clearly very difficult to tell if a habitat action has had a positive effect on a population if downstream that population has been negatively impacted by dams, warm water, predation, fisheries, or even further afield by ocean conditions.

Restorationists often refer to the ocean as a "black box." Research on the ocean phase of the salmon life cycle has been scarce compared to research on the freshwater phase. This is partly due to the fact that most restoration and mitigation actions occur in the rivers. But this research focus is also due to the legacy of the scientific institutions and organizations that were created in the early 1900s. Hatchery research and "counting fish" has been the status quo for decades (Taylor, 1999). Counting "fish in and fish out" of the river system has been one of the only continuously collected data since Harlan Holmes's time. Slowly, this numbers focus is beginning to shift as the ocean becomes recognized as a major factor in salmon abundance—partly due to the effects of the phenomenon of the "blob" described in previous chapters. Ocean life cycles are now coming to be viewed as a critical factor in salmon recovery, and new metrics for restoration success that take larger spatial and temporal scales into account are emerging as ecological restorationists themselves also shift their practices to meet the challenges of climate change. According to one practitioner, because of these challenges, "you are never really going to have an understanding of what you did right and what you did wrong." The long-term project of restoration lends itself to long-term monitoring, yet the resources are often misaligned with this need.

While the large scale and complexity of fish recovery exacerbate the problem of monitoring, there is still a need to understand whether or not the millions of dollars being spent on habitat restoration is working. Many people still question whether monitoring is an appropriate use of funding; yet most restorationists understand the importance of collecting monitoring data and establishing baselines. They also advocate for rigorous scientific work to back up their contention that the habitat restoration they are doing is actually "moving the dial" in the right direction. Despite this, most restorationists also agree that establishing a comprehensive monitoring program presents major challenges. One state-level restoration policy-maker said: "It has been challenging to fund, and it has been challenging to coordinate, and it has been challenging to outreach the results."

Although, on the surface, it may seem like a lot of money is going into monitoring—and "research monitoring and evaluation" accounted for

30 percent of BPA's recovery budget in 2016—only 16 percent of that portion, or $13.3 million USD, went toward *habitat* monitoring (NW Council, 2017). The majority of the monitoring money actually goes toward hatchery effectiveness evaluation, *not* restoration monitoring (NW Council, 2017). In a technologically updated version of Holmes' marking research programs of the 1920s, the most common monitoring action consists of PIT tag arrays, which count tagged fish that pass by them on their way upstream or down. Yet most PIT tags are in the flesh of hatchery fish, not wild fish. Of course, even then, this data is able to provide only abundance estimates that can be used in models, not absolute population numbers. While detecting numbers of fish is crucial to figuring out if recovery targets have been met, counting fish does not necessarily demonstrate the effectiveness of habitat restoration itself—especially when the difference between achieving or failing to reach a recovery goal can be as little as a few dozen fish.

Even if it seems that a lot of money and effort is being spent on monitoring, people are anxious that it might not be going toward the right things or that it is simply not enough. For instance, although some BPA mitigation funding is funneled toward monitoring through the ISEMP, CHaMP, and IMW projects, there are many other local projects that only see minimal monitoring, usually enough to meet BPA's required five-year period. This means that even large restoration projects are often implemented with almost no effectiveness monitoring to follow up on the results of the project. For some restorationists, losing this chance to collect data and understand ecological systems is a grave loss to science. As one restoration contractor lamented to me in his firm's office, which was brimming with blueprints as they prepared to launch a new large-scale restoration project:

> Nobody is paying for monitoring and nobody is paying to publish the lessons . . . We are doing some of the biggest floodplain projects in the western United States, yet I can't point you to a good case study about them because nobody wants to spend the money to write it up. You have probably never even heard of the [recent, large] restoration that we have done. It's been a fourteen-year project. They spent millions. Millions and millions of dollars have been spent and it's revitalizing critical blueback salmon habitat.

And it is working! It is really enlivening a two-mile floodplain. Holy moly! It's incredible! I could go on all night about the projects we have done that are really changing the local ecology and changing how floodplain function is benefiting salmon and I can't give you a single case study. To write up one of those case studies is thousands and thousands of dollars . . . but can we spend 1 percent for science? Can we spend 2 percent?

Finally, while restorationists want to know if their work is making a difference, there is also anxiety about what the answer will be. For example, one tribal restoration manager worried:

The IMW gives us the opportunity to put all of our money in one place and say: "Okay, what if we *did* have enough money to provide restoration treatments to all of the areas that we think are suitable for restoration? Is that enough?" And I think a lot of people are really afraid of that answer. Because what if it isn't? What if we take all these areas on [public] land where we have a supportive landowner and a mandate to protect and we do all of that and it's not enough? That's a really ugly spot to be in. Right now, with the money that we are putting into [restoration] we should be able to restore the Columbia in about two hundred years, and that assumes no further degradation. So, if I go in front of [the funding agent] and say, "We are doing great work and all you have to do is keep funding us for the next two hundred years, and we are totally going to get to our recovery goal!" What then?

Indeed, what if restoration monitoring demonstrates that it has been ineffective? What then? Some of those who I spoke to fear this is a very real possibility, given the fact that climate change is already heavily impacting the basin some areas.

ACCLIMATION: MONITORING FOR THE FUTURE

"Acclimation," in an ecological sense, is a process of adjustment to a new environment that occurs over a relatively short period of time in an individual. Examples include aquatic species adjusting to changes in water PH, or an animal shedding its fur in the spring as the weather warms. This is in contrast to "adaptation," which occurs over a generational timescale. Because I am discussing how individual restorationists are in the face of uncertainty changing their practices within a short timescale and making decisions immediately, acclimation provides an appropriate metaphor. Acclimation is about choice and action. It is about how people act in the face of uncertainty.

Yet, acclimation as an adaptive move is different from adaptive management: It is a way of doing daily and detailed epistemic work that embraces change and uncertainty in the immediate present so that action in the field can be taken. In contrast to adaptive management, it occurs at a cultural and practical level on a day-to-day basis. These gradual shifts in practices are a process of adjustment, yet they take place over a relatively short time period. In many cases in the Columbia River Basin, there is no time to wait for answers about what to do or what to monitor: inaction is unacceptable because without something being done species will go extinct. Yet instead of ignoring or putting off making decisions about which actions are effective and what to monitor now, restorationists are instead moving forward in their scientific and management work and developing monitoring strategies that allow them to adjust.

While restoration and monitoring programs have been fragmented across the basin making coordination difficult, a basin-wide monitoring coordination effort is still being created. This is an organized attempt to meet the needs of monitoring in all restoration programs across the basin and for restorationists to orient their monitoring toward the future. The development of the Pacific Northwest Aquatic Monitoring Partnership (PNAMP) provides an excellent example illustrating how the concept of acclimation is employed when dealing with future uncertainty and the need to monitor in the present.

PNAMP was created in 2004 to coordinate among federal, state, tribal, and private organizations involved in aquatic monitoring programs throughout the Pacific Northwest. The partnership's main goal is to work as facilitators to help coordinate monitoring efforts efficiently and effectively across the basin so that those efforts provide information to inform decision-making at multiple scales. PNAMP aims to accomplish this by facilitating the exchange of scientifically based monitoring data collected by voluntary partners throughout the region. PNAMP has several ongoing projects in which they are working to standardize and coordinate aquatic habitat monitoring throughout the region. One of these is the Integrated Status and Trend Monitoring (ISTM) Project. The ISTM is a strategic program that aims to demonstrate how coordination and standardization of habitat data can help address questions that bridge local to region-wide scales (Puls, Dunn, & Hudson, 2014).

Another PNAMP program, the Coordinated Assessment project (comanaged with the PSMFC StreamNet Project), is aimed at standardizing data on anadromous fish so that it can be shared throughout the Columbia River Basin. Through ongoing working groups that span multiple agencies, PNAMP partners have been developing a list of "regional habitat indicators" that can be used across the entire basin. PNAMP facilitates data standardization through web-based tools, standardized data collection protocols, and knowledge exchange. One PNAMP working group is tackling this standardization project in anticipation of climate change and is developing a list of standardized measures in order to establish ecological baselines *now*, which will still be useful in a climate-changed *future*. Deciding what to measure *now* requires thinking to the future and designing knowledge infrastructures that take the "long-now" into account (Ribes & Finholt, 2009). This long-now could include changes in policy, technology, or the environment, and involves adjusting the ways that data is standardized, classified, and collected.

CHANGING DATA COLLECTION METHODS

Technoscientific change, or changes in science more generally, often occur when new innovations or technologies become available (Ribes, 2014)—and this is no different in the Columbia River Basin, where the ways in which some ecological data are collected and recorded have shifted relatively quickly over the past two decades. Digital technologies have replaced analogue field methods, and increasingly sophisticated remote-sensing capabilities have become more accessible. Emerging technologies are facilitating changes in small- and large-scale monitoring. New monitoring instruments and technologies include inexpensive temperature logging devices, as well as more expensive drones with lidar (a surveying method using lasers that can show detailed topography) capabilities. Changes also include the development of more sophisticated and detailed computer models to process this data, often using satellite uplink systems for instantaneous field data collection.

Yet even if the data collection practices are changing, many of the objects that are being measured are the same. In addition to leading the PNAMP coordination project, Jennifer Bayer has been a fisheries biologist for over two decades. She is clearly passionate about not only restoration science but also the collaboration opportunities that she is forging, and she agrees: "I think that there's a big shift in how we collect data and process and handle it. But as far as *what* we are collecting, it still feels like it is largely the same." In other words, regardless of what is being measured, the way it is being measured, and the way the data is being handled and modeled, is changing. New monitoring practices are helping restorationists adapt to environmental changes, including the need to understand environmental variables at different temporal and spatial scales and in greater detail. For example, temperature gradients within streams can be measured with detail and accuracy using temperature loggers, but drones can now use forward-looking infrared (FLIR) technology to identify cold-water refuges and "seeps" or springs within streams. These areas might have been overlooked in the past; but now that maintaining stable, cool, water temperature is so critical for salmon that are facing climate change, these tools are providing essential information. Until recently, one restorationist recalled, "a lot of questions couldn't be answered at the thirty meter–scale range, but now it [that data] is available." These new, more detailed data-collection practices are becoming integral to the monitoring effort and are essential to understanding whether or not restoration is working, what actions to take, and what kinds of metrics might be available or useful in the future. This is a key part of PNAMP's work and an example of scientists continuing to act in the face of uncertainty through pragmatic efforts in which they are developing new data and metrics and embracing new technologies.

TRAINED JUDGMENT

Covering restoration sites with temperature loggers and groundwater monitors is one way to prove that restoration is positively affecting stream temperature. Setting up nearby watersheds or upstream reaches as controls where there hasn't been any restoration is another way; but this work is time consuming, expensive, and might not provide answers fast enough to enable adaptation to climate change effects. While rigorous experimentation and data collection is still the most respected form of creating new knowledge in the restoration community, shorter-term observations are increasingly important. Recalling the history of temperature monitoring, one regional restoration manager reminisced over his thirty-year career:

> Fifteen or twenty years ago it was a lot of guesswork. In the last decade, we have reduced a lot of the guessing into knowing. What we do is a relatively new industry. There is a lot of learning going on. Some of the stuff we are going to do we know is effective. We won't know *how* effective because every individual site varies—but we know that if we store more water in the ground, that we are going to have a positive impact on stream temperatures and flows.

Opinions like this one are common: if restorationists can *prove* that what they are doing is working, that is, of course, better. But if not: many of them still expressed a belief that the actions they are taking are making positive difference in adapting to climate change. Another landscape-scale restoration researcher pointed out, "If we can't get the baseline, maybe we should be looking at the trend . . . I'm a proponent of learning by doing. Sometimes just starting to do something allows you to learn something, which allows you to do it better."

Elinor Ostrom (2009) highlighted the importance of measurability of environmental variables in socio-ecological systems. In an adaptive socio-ecological system, finding ecological markers that can be measured accurately and reliably enables mangers to adapt to changes, whereas unpredictable or invisible conditions hinder the learning that is necessary to adjust management decisions (Ostrom, 2009). Yet restorationists not only acknowledge the need to act with imperfect information, they also recognize that this is often the only way they can move forward in their work. These kinds of practices now permeate restoration work as acclimation is being embraced with the onset of climate change. As one restorationist pointed out to me during a workshop on monitoring: "empirical information is preferred, but sometimes we need to connect the dots between what we did and how the fish respond." In the end, being able to say *something* about trends and patterns over the long term is the best that many can hope for.

While long-term, rigorous monitoring programs are clearly important in the broader scope of species recovery, more immediate information is also required. Restorationists need to know whether or not their riparian planting or floodplain reconnection project is making a difference in lowering stream temperatures because salmon survival may immediately depend on it. They also need to know if their engineered log jams and channels are withstanding flooding events and behaving the way they expected. To answer these questions, many restorationists have decided to simply use their own expert interpretation, or what Daston and Galison (2007) call "trained judgment." This ongoing work allows present epistemic needs to be met, while facilitating constant adjustment to a changing environment.

While some restorationists would rather (or are required to) monitor to prove and quantify the impacts of their projects, many restorationists have decided to simply use their own "trained judgment" as a more common method for drawing inference. In order to deal with gaps in information, some restorationists have worked quickly to come up with their own monitoring programs, many of which have minimal monitoring protocols but are complimented by the trained judgment of experts in the field. One project manager recalled the process as leaving them frazzled:

> We cobbled together some funds to jury-rig together a monitoring plan . . . we had spawning survey crews going through and we just kind of by hook or by crook got this monitoring together. We were spending a lot of money and we were putting a lot of work in on the ground and we really didn't know what was going on.

These kinds of "quick and dirty" monitoring programs are commonplace and illustrate a local-scale adaptation in scientific practice that uses the resources at hand, as well as trained judgment as needed.

One woman, who had worked on monitoring projects for decades, found that going back to "the old science" and "just going on a good old-fashioned walk" around a restoration site was a useful practice. Scientists that I spoke to often described field walks such as these as a kind of monitoring practice. Walking throughout the site, they could observe where structures withstood high water or where groundwater was infiltrating the floodplain. According to another conservation biologist,

> Those [observations] are extremely helpful . . . I don't need to be a statistician. All I need to say is, "This project is designed like this, and was intended to do this. And, this is what it looks like once it was finished, five years later or seven years later" . . . what I need to know is how did this design element, in this hydrologic, geomorphic situation, affect the fish habitats. That is what I'm after.

Because it lacks an experimental design, this type of observation may not be deemed "monitoring" by many restorationists, yet it is nonetheless all that many can manage: due to financial and time constraints, these kinds of on-the-ground observations may be the only monitoring activity that some restoration sites receive. In the absence of a formal field study, many consider those judgments to be sound enough to warrant decision-making. Practices like these, which incorporate the trained judgment of individual scientists are therefore helping to verify larger-scale assumptions and inferences. These inferences are then useful to other diverse ecological circumstances throughout the basin, all of which are anticipating the localized effects of climate change in different ways. These observational strategies are allowing restorationists to develop new baselines, categories, and monitoring protocols that will remain useful into the future. Making trained judgments such as these is therefore an adaptation that lets scientists make epistemic choices in the face of uncertainty, helping them acclimate to climate change—environmental flux which they are currently witnessing in their experimental data and monitoring, and in their daily observations. While this kind of scientific practice may not be ideal, it is a way forward, and it can be seen in the monitoring strategies that many restorationists end up using in the field.

CHANGING METRICS: STANDARDS FOR THE FUTURE

Another way that restorationists are orienting their work to alternative baselines is by establishing metrics that will remain useful in a changed climate. Metrics for understanding change in novel ecosystems include "future habitat condition," which allows model projections to be incorporated; "functional diversity," which recognizes resilience in an ecosystem (Suding & Leger, 2012); and even the valuation of different aspects of an ecosystem through measures of biodiversity (Ankersen and Regan, 2010) or ecosystem services (Evers et al., 2018). Restorationists are recognizing the need for different benchmarks for success, or what one restorationist referred to as ways of "seeing through metrics." For example, as water and air temperatures increase, lowering stream temperature is emerging as an important metric for success, and creating shade through vegetative planting and connecting streams to cooling ground water is now becoming a focus of restoration design. In response, doing more frequent water temperature monitoring and setting benchmarks for temperatures at restoration sites is now common practice. As new information emerges about the kinds of restoration actions that mitigate climate change effects, new metrics for monitoring are thus being developed by restorationists as well as policy-makers.

Yet the intensity of co-production between science and policy also leads some restorationists to refer to restoration science in the Columbia River Basin as an "ESA-centric science," whereby fish populations are oversimplified—complexities within species are "flattened" into simple categories and success is simply counted using population numbers. Some fisheries biologists view this categorization and standardization of fish populations as a problem that is at odds with diversity, a quality that is becoming recognized as critical to adaptive capacity. Indeed, the ESA often seems to facilitate a science that counts fish. This is partly because the biological recovery goals are often numerical ones and set for each population and species using biological opinions and recovery plans. This quantification and categorization of fish in terms of run-specific populations has not been straightforward. For instance, drawing a distinction between spring and fall Chinook can be problematic, especially when those life histories and run-timings are shifting with climate change. As one fisheries biologist, troubled by the way that salmonids are categorized, pointed out: "We love to categorize them. We love to throw calendar dates on them. All of that is just detrimental to respecting diversity. There are numerous populations and they don't fit into the norms and we just ignore that."

Fisheries biologists recognize this diversity as being a critical feature that contributes to the viability of a population because it gives a species the capacity to adapt to environmental changes. Yet, the ESA as an institution requires metrics and measures which align with BiOps and recovery plans but not necessarily with biodiversity. Some worry that a focus on the ESA's limited metrics could stifle salmon species' adaptive capacity. As one policy-maker complained: "We get all wrapped around the axel on the ESA, but it can also be very detrimental."

From the standpoint of restoration science, the ESA and the legal requirements for recovery entail a certain kind of applied scientific lens, which can come to dominate monitoring efforts especially through the recovery plans and BiOps. Another restoration planner worried that:

> It forces us to look at the world through a certain lens and if the biological opinion says over the next ten years you have to deliver X number of SBUs [survival benefit units] then that is what we are judged against. That forecloses the opportunity to maybe do other things that would be yielding different benefits that may be harder to quantify or that occur over a longer time frame.

Some restorationists fear that by limiting their focus to certain metrics other opportunities for restoration could be missed. Restorationists like Bayer at PNAMP are also cautious about altering these kinds of measures and acknowledge that the rigidity of the ESA is also what makes it a strong environmental law. This tension between flexibility and rigidity in law goes deep into the science of restoration itself, and it is always under the surface.

While fish numbers can be counted and estimated fairly well, making a judgment about whether or not *habitat* is improving has not yet been possible given the data that is available. This means that recovery is understood by using the data at hand. While we could conclude that the ESA and other regulatory frameworks are *dictating* the knowledge that is needed in the Columbia River Basin, co-production provides a more dynamic perspective whereby science can also drive policy. In employing different metrics, such as the presence of macroinvertebrates, restoration scientists in the Columbia River Basin push back at the metrics mandated by the ESA, finding new ways to measure recovery success that might be more appropriate than simply "counting fish."

Just as measuring the success of a beaver restoration is difficult, many of these new metrics are problematic to operationalize. For instance, it is difficult to create metrics for some ecosystem attributes because those attributes are often difficult to "see" through metrics. Measuring the length and geomorphology of an anastomose, or complex, braided stream is difficult and time consuming, and no standard metrics have been developed to do so.

Similarly, functional diversity is highly complex and difficult to quantify, yet this indicator may be one of the most important in terms of climate change, as ecosystems that have a mixture of species serving functionally diverse roles may be more able to mitigate extremes of temporal variability (Suding and Leger, 2012).

PNAMP is attempting to create a knowledge infrastructure that, while oriented to the future, is still useful today. In order for a knowledge infrastructure to function, it relies on standardization and classification (Star & Bowker, 2010). And, although the process is often overlooked, setting standards is a scientific practice that garners a lot of time and energy. According to science studies scholars, Susan Leigh Star and Geoffrey Bowker (2010), knowledge infrastructures are actually made up of multiple layers of standards. In their words: "it is standards all the way down" (Star & Bowker, 2010, p. 6). While standardization and classification have political implications—because "seeing" something and counting it can be matters of power and control (Scott, 1999)—standards also have practical implications for the everyday work that scientists do. Standardization allows for reproducibility of data and, by providing clear metadata, information can be both be shared and reused. This is important because it allows scientists throughout the basin to use varying data-collection methods yet still communicate across the region. PNAMP's facilitation of this process is therefore an important, and often overlooked, step in creating a useful knowledge infrastructure.

Bowker and Star (2000) differentiate between standards and classifications although they are closely related. They argue that classification systems are a central feature of social life (Bowker & Star, 2000). Classification is a complete system with consistent segmentation, in which everything has a place, whereas standards are sets of "agreed-upon rules" that "span more than one community of practice" and "persist over time" (Bowker & Star, 2000, pp. 13–14). This allows standards to work across spatial and temporal distances. Standards are often enforced and even created by legal bodies or set in order to satisfy legal requirements; and they also possess their own "inertia," because they are difficult to change (Bowker & Star, 2000). This inertia is an important point to consider in the Columbia River Basin, where restoration is often driven by the ESA and the metrics that the act requires.

The work that PNAMP is doing to create regional habitat indicators provides a good example of the "practical politics of classifying and standardizing" (Bowker & Star, 2000, p. 44). In other words, this is the often messy and heavily negotiated design work that goes into creating a knowledge infrastructure. One working group member described the process this way:

> We thought that the coordinated assessment project would move a lot faster. . . . But it turns out that it is a lot more time-intensive and complicated to get the indicators that we want . . . every year is different, which means that all the data compilers have to go out and work with a biologist and say, "Well what is your analysis saying this time?" There is a lot of documentation and work that goes into getting that data.

In developing the regional habitat indicators, the partners involved in PNAMP have worked through a lengthy process of negotiation in which trade-offs between different standards were clarified. After this process, the working group decided on a set of indicators that they believed could demonstrate changes in the environment relevant to salmon recovery.

These indicators included: flow, macroinvertebrates, stream temperature, and water quality. Questions concerning what could be measured and how it could be measured most efficiently often arose during discussions. These practical difficulties were especially clear when discussing the newest indicator: macroinvertebrates. Part of the PNAMP working group's reasoning for looking to a new indicator such as macroinvertebrates was to supplement the old data paradigm of simply counting fish. As Bayer put it, counting numbers of fish is not the only data needed to understand whether or not salmon are recovering. By counting something new—macroinvertebrates—"We are trying to gently say: 'you can count other things.' Because there are a lot of things that affect salmon, because it is many years before they come back!"

While water quality, temperature, and flow have been measured with regularity in most parts of the basin for decades, macroinvertebrates have not previously been extensively counted by fish and wildlife managers. Therefore, the kinds of data that are needed to understand this indicator are not as readily available. While there are some water quality assessments that use benthic macroinvertebrates as a measure of water quality, an infrastructure that supports this data collection to assess habitat is not yet in place, nor is it agreed that macroinvertebrates are the right indicator for habitat assessment. Despite this, the group acknowledged the potential importance of the indicator and the ways in which it could become useful in the future. Setting aside the practicalities of collecting macroinvertebrate data across the basin right now, the working group decided to keep the indicator on the short list just in case it could be useful someday, when collecting macroinvertebrate data in all streams may become more practical. By thinking creatively about future epistemic needs, the working group members were able to come to agreement and adjust their current work accordingly.

Through the work of the Regional Habitat Indicator Project, it became clear to the participants that creating a comprehensive data collection system that was completely standardized was going to be an impossibility. Yet throughout the negotiation process, participants realized that complete standardization actually may not be entirely desirable. As Bowker and Star (2000) point out, residual categories, or those that don't quite "fit" into the classification system, do often provide an important way to "dilute uncertainty." Restorationists also see that keeping these "residual categories," which don't quite fit into a classification system, is an important hedge against climate change. In the words of one working group participant: "Don't put all your eggs in one basket. Measure multiple things." The residual category of macroinvertebrates might not entirely fit into the system of standards at this time: but instead of discarding it, keeping the category and setting it aside to be used in the future builds in the potential to adapt to future knowledge needs and is another example of "acclimation" in action.

Once a recommendation like the regional habitat indicators is in place, this kind of standardization and design work is often overlooked (Bowker & Star, 2000), but it remains an important part of the epistemic work that ecological restorationists in the basin are constantly undertaking. As one participant in the working group said:

> It's the tough part. It's really contentious and it's really hard to do. You end up coming up with a lot of surrogates and for practical reasons there is an ideal indicator that is actually possible given your budget constraints. So, there is a real art to it. When it is done well it is incredibly powerful and galvanizing, and you can get a diverse set of people around a table to accomplish something much greater than any one organization can do.

Standardization work must correlate across large stretches of time as well as space. Fish lives, BiOps, careers, and climate change must all be considered. As new information about what works and what doesn't in restoration becomes available, new metrics for monitoring may need to be developed. Quantifying diversity of habitat, Stage 0 streams, or macroinvertebrates is undoubtedly difficult; yet these indicators may be some of the most important ones when it comes to understanding the effects of climate change on salmon and their environment.

By thinking pragmatically and creatively to imagine future epistemic needs, restorationists are "acclimating" their knowledge infrastructure to the changing climate. In terms of designing a sustainable infrastructure, this requires considering future concerns, or the "long now" of infrastructure design (Ribes & Finholt, 2009). By looking to these long-term temporal and large spatial scales, infrastructure developers can intentionally incorporate management goals and desirable futures for the Columbia River Basin into infrastructure development. In this way, designing for the long now in infrastructures becomes a potential adaptive strategy, as it considers long-term sustainability and the need for adaptive capacity (Ribes & Finholt, 2009).

RETHINKING MONITORING FOR THE FUTURE

Restoration monitoring has always been complex and contentious. Yet, as climate change is making itself felt, monitoring in the basin is nevertheless moving forward, even with the additional complexity that is being introduced. In a sense, restorationists are being forced to acclimate. This is a process of adjustment—an acclimation of epistemic work in which choices are made about how to move forward. Acclimation describes how restorationists can act in the face of uncertainty, adapting their practices and the knowledge infrastructures that support them. Along with emergence, these strategies are helping the epistemic community deal with and adapt to change. Looking to how knowledge infrastructures may remain flexible and adaptive is an important step in facilitating this acclimation to climate change in scientific work.

Understanding how knowledge infrastructures are changing has been identified as one of the key research challenges for infrastructure studies (Edwards et al., 2013). Ideally, knowledge infrastructures should support scientific work without needing constant upkeep—therefore they should be relatively stable and reliable (Star & Ruhleder, 1994). However, infrastructures are also "paradoxical" in the ways that they both support and stifle adaptation and change (Star & Ruhleder, 1994). According to Bowker and Star (2000), this is due to the way they must be able to facilitate work practices across organizations and users by employing standards, while at the same time remaining locally useful and specific. This means that knowledge infrastructures need to be both rigid and flexible: universal yet able to change (Edwards et al., 2013).

This tension becomes especially clear in large-scale infrastructures where sociotechnical systems have a spatially and temporally broad reach (Star & Ruhleder, 1994). While infrastructures are often understood as "geographically dispersed" (Star & Ruhleder, 1994, p. 112), they can nevertheless be built to facilitate the management of a natural resource and promote a specific management paradigm and goal in a specific location. Shared infrastructures are an important "public good" that are ideally oriented toward supporting sustainable research over the long term (Bowker et al., 2010). While the locally situated, yet larger-scale, dispersed nature of infrastructures makes them particularly complex, at the same time this is what makes them more likely to be able to adapt, as they have multiple points where adaptation can occur and more resources to draw from.

This chapter has brought attention to the ways in which knowledge infrastructures, including data collection methods, metrics, and standards, must also adapt to changes, including both technoscientific and environmental changes. What I have found through this study of the restoration community is that they are acclimating these infrastructures through a process of adjustment: both formally, through groups like PNAMP, and informally, through the practices of individuals and their use of trained judgment. This demonstrates the adaptive thinking that is possible in the face of uncertainty, even within an epistemic community like ecological restoration, which has been so keenly focused on historic baselines and monitoring to meet requirements often dictated by the past, or by legal mandates such as the ESA.

6

Anticipation

Imagining a Future

IN MANY SCIENTIFIC fields—especially ecology—computer modeling has come to "complement or even replace" both laboratory and field experiments (Edwards, 2010, p. xix). In ecological science, models are used to organize data, synthesize information, and predict the future (Oreskes, 2003). Information within the Columbia River Basin is fast becoming model-dependent: either being "fed" into models or being derived from them. To support this work, a modeling infrastructure has emerged in the basin. These models are often used for decision-making, and they are becoming more complex as they account for an increasing number of environmental and social variables. Unfortunately, this complexity can often be at odds with the usefulness and accuracy of models when employing them for prediction or decision-making (Oreskes, 2003).

Models can be used to understand how multiple and overlapping scales relate and interact. Chances are, data that is collected at a local field site are also being fed into a model to understand larger-scale dynamics. In the Columbia River Basin, modeling for climate change is accomplished using global climate models, in addition to down-scaled and localized models that focus on one or two variables such as stream temperature or sea-level rise. These more local-scale models can help facilitate monitoring, planning, and prioritization of restoration sites. Local-scale modeling efforts in the basin also include fisheries life-cycle and population models as well as hydrological models related to dam operations or stream flow. As early as 2007, models were being used to predict climate change impacts to salmon and habitat restoration itself in the Pacific Northwest (Battin et al., 2007). While the predicted "large negative impacts of climate change" (p. 6720) have proved themselves true, modelers in the region continue to adjust their work to make finer and more accurate models and predictions.

In addition to the changes in data collection and the use of trained judgment discussed in the previous chapter, many restorationists rely on models to fill in "gaps" in information or to understand climate, hydrologic, or ecosystem dynamics in areas where monitoring does not occur. In the past, predictive decision-making tools such as models were based on historical observations and baselines, yet now decisions need to be made in an increasingly dynamic system, sometimes with drastically shifting baselines. Monitoring and modeling are inextricably linked: monitoring data must be continuously collected and then used to update and validate models themselves, ensuring that they are adaptive and useful for planning or prioritization. Yet, while the large-scale impacts of climate change on salmon and their habitats are relatively well understood, at least in terrestrial waters, the responses of communities and individuals at finer spatial and temporal scales are not as easy to model, and so are far less clear (Conroy et al., 2011). Further, the combined impact and increased variability in climate in addition to interactions among different dynamics such as changes in snowpack and wildfires, and shifts to longer, dryer summers, are just beginning to be understood. Adaptive learning through modeling is therefore a strategy that restorationists are using to overcome these gaps in knowledge.

"Anticipation" is an affective state of "thinking and living toward the future," and it can permeate scientific practice, becoming an epistemic virtue in itself (Adams, Murphy, & Clarke, 2009). In an effort to restore to historical ecosystem states, ecological restorationists have often oriented their work toward the past. With the impending (and present) impacts of climate change, however, this is no longer possible. Restorationists are, instead, more often orienting their work toward the future and aligning scientific practices and knowledge infrastructures toward potential futures. The emerging virtue of anticipation, as well as other forms of "anticipation work" (Steinhardt & Jackson, 2015) are therefore key strategies that ecological restorationists are using to adapt. I am calling this strategy "anticipation," and it can be illustrated through the way in which the scientific practice of ecological restoration has shifted from being mainly fieldwork-based toward becoming increasingly modeling-centered. This shift in the locus of knowledge production has occurred at a basin-wide, institutional scale, as well as in the practices of individuals. While field-collected data still informs modeling, new models are being developed that are taking the place of some field-produced knowledge. Restorationists are coming to rely less and less on field science alone. Instead of gathering data about current and past conditions and using them to draw a baseline and determine restoration goals, the future is often simulated through predictive modeling and the river is being restored in light of this anticipated future.

MODELING AND UNCERTAINTY

Very broadly conceived, modeling is a tactic for dealing with uncertainty in science (Conroy et al., 2011). Climate change itself is understood through models (Edwards, 2010). Yet, within models themselves different types of uncertainties necessarily already exist. Many of the monitoring issues outlined in the previous chapter are related to the need to establish a baseline from which to measure changes in the environment. Yet establishing a good baseline in a highly altered system like the Columbia River Basin is a major challenge. Further, dealing with a baseline that is constantly shifting as the climate changes makes "baselining" a highly complex task (Ureta, 2017). Useful baselines are just beginning to emerge from the past two decades of restoration monitoring work; but these baselines, which are often made through intensive monitoring on a single stretch of stream, can also be destroyed by an unexpected turn of events such as a fire, drought, or high river flows.

While most of the uncertainties inherent in models can be accounted for within the modeling process itself, others are more fundamental to socio-ecological systems. One reason modeling is such an important element of any adaptive scientific practice is that it is a way of quantifying uncertainty. Uncertainty in environmental modeling can be classified into four types (Conroy et al., 2011). *Environmental variation* includes changes in circumstantial variables, such as climate. In the Columbia River Basin, these changes are now mostly assumed. Although this effort is complicated by the increasing frequency and intensity of environmental variations, environmental modelers are dealing with these uncertainties by incorporating downscaled climate modeling as well as other modeling tactics (Abatzoglou, 2013).

Because of the large and diverse spatial scale through which salmon migrate over the course of their lives, the second type of uncertainty: *partial observability,* is still a major issue in the models that restorationists use. Partial observability, or the need to estimate what is unseen, establishes the need to estimate through modeling in the first place (Conroy et al., 2011). For example, many restorationists that I spoke to refer to the ocean as a black box, where little is known about what salmon do, where they go, or what their survivability rates are in varying ocean conditions. Early models simply included a best guess at survival rates, and models were run using that number. Yet, as more data become available and the life-cycle models are refined, survivability is becoming more calculable, and uncertainty more quantified.

A third type of uncertainty, *partial controllability,* is also being exacerbated by climate change, and is making modeling an even more important tool for understanding the environment. Partial controllability is the inability to apply management actions or experimental controls precisely across sites (Conroy et al., 2011). Designing experimental controls in a highly stochastic system such as a river is often practically impossible. Drawing inference between watersheds thus becomes difficult without the use of models, and partial controllability becomes one of the fundamental factors contributing to uncertainty in any model of a river system.

Finally, *structural uncertainty* describes the uncertainty inherent in models themselves. While the first three kinds of uncertainty can be dealt with through statistical distribution, the last one—structural uncertainty—cannot be dealt with statistically. It will always be present and needs to be clearly stated in order for a model to be useful for decision-making (Conroy et al., 2011). By clearly stating the uncertainty inherent within their models, modelers can help restorationists deal with these fundamental uncertainties.

Although modelers must deal with all of these uncertainties, models themselves remain essential tools for addressing uncertainties. Most importantly, models help restorationists anticipate the future by creating metaphors and tackling difficulties associated with complex cross-scalar interactions and epistemic indeterminacy.

In the following section, I will introduce two modeling practices that are anticipatory in nature: a large-scale stream temperature modeling effort and life-cycle modeling. These examples will allow me to describe some of the anticipatory practices that restorationists are using. It is important to remember, however, that these modeling practices must also be supported by an adaptive knowledge infrastructure that anticipates the future by allowing scientists to deal with complex scales and helps them tame indeterminacy. Finally, I describe the ways in which modeling allows restorationists to anticipate and envision the future.

A MODEL OF THE FUTURE RIVER

NorWeST StreamTemp, a project facilitated by the US Forest Service, has gathered together stream temperature data that were recorded by over one hundred resource agencies across the western US. These data were then compiled into spatial statistical network models using air temperature and discharge rates from thirty-six historical and future climate scenarios. The result is an interactive temperature map that shows current and expected temperatures for all streams in the western US.

Creating the database was no easy task. According to the project's leader, Dan Isaak, "it is a long and sordid story." Over the past decade, Isaak has been working to bring together disparate stream temperature and water quality data from across the basin. The project has a much longer history, however, and emerged from work he saw being conducted as a graduate student in the late 1990s:

> I was actually picking up a project that one of my predecessor scientists had worked on and then dropped . . . they pulled together

a big database from biologists across the interior Columbia, and . . . for some reason they compiled that dataset but they never published it. This was about 1998. That sat on his shelf moldering and gathering dust.

Yet something about that early work stuck with Isaak, and he ended up returning to it decades later. As a fisheries biologist working in the Intermountain West, he remembers climate change getting some attention even in those early days. Yet somehow, he says, it feels like it got set aside until more recently:

For me, climate change had been an interesting topic because when I was back at University of Wyoming one of the professors where I was doing my PhD had done some of the early work. He was making these big West-wide projections about trout basically going extinct because they were going to lose so much habitat. And I always thought that was kind of interesting. Then, when those papers came out by Phil Mote describing that climate change is something that we have been living through essentially for the last 50 years . . . I was like, "Wow this is actually not a hypothetical, this is a real thing."

While the scientific evidence of climate change effects were emerging, political shifts facilitated the funding to get the research off the ground:

At the same time, Al Gore's movie was coming out and people were getting kind of amped up about climate change. We had a few hot summers in the Northwest and a lot of fires and people started saying, "Oh, this is something that we really need to be focused on." That led to grant opportunities to study it further.

When Isaak talks about the StreamTemp project, his excitement of the challenge is palpable. It has been a difficult and time-consuming task, so in order for it to get done, the project needed someone who would be motivated enough to even attempt such a large project. Although the data existed, they came from hundreds of sources throughout the basin and were often not interoperable. While scientists like Harlan Holmes collected data on fish numbers and sometimes even stream temperatures, much like restorationists in the Columbia River Basin do today, the data that were collected in the early twentieth century first had to be contextualized to current standards and locations. Similarly, stream temperature data from the past several decades have been collected, but in order for it to become useable, the data had to be cleaned of errors. Transitioning the data to the database required a team of full-time staff to "clean" it:

> The data was just sitting around scattered on people's different hard drives. . . . Usually we had two or three people working full time on the data team just getting data files, cleaning them, and organizing them . . . so the hard part was just grunting and mucking through that much messy data . . . luckily one of the good things about temperature data files is that they are all pretty much in the same format . . . they give you a date and a time stamp and a temperature. But you still have to think about where specifically the data was collected, and a lot of it was collected pre-GPS days, so they didn't have good coordinates. We'd be on the phone with the people that sent us the data and try to figure out where it was coming from.

The temperature data was unique in that much of it was already interoperable; but even so, the task of "cleaning" the data was one of the major hurdles to creating the model.

Nevertheless, in terms of their utility to restorationists, the maps derived from these models have become invaluable in determining where restoration may have the highest impact, especially for cold-water species of bull trout, a species of salmonid that is particularly sensitive to high temperatures, dictating their survival and spawning patterns. While some restorationists that I spoke to were concerned that the ability to prioritize and locate streams that may be too hot to support salmonids in the future could lead to "giving up" on restoring some areas, Isaak points out that the models are not intended to be at a scale that would be precise enough to enable this kind of decision-making. Instead, they are meant to help restorationists identify environmental trends and general areas in which to focus restoration efforts. Instead of acting as a "triage" where only streams that show colder waters in the future would be restored, these models can also have the opposite effect—focusing restoration efforts on areas that need them most. In terms of the region where Isaak usually works, this often comes down to the fly-fishing industry:

> In the case of Montana, which is a world-famous place and a mecca of fly-fishing, we can now show the rivers that are right on the edge of becoming too warm . . . the Madison, the Yellowstone, the Bitterroot, some of the places on the fly-fishers' bucket list . . . and say, these are not going to be very good trout streams in the future unless you go in and do habitat restoration and try to cool things down . . . if we can pinpoint some of those places that people care a lot about, and maybe if they do something there, they could make a big difference.

But the models are not only intended to support the recreational fishing industry; restorationists across the basin are excited about the potential for using these models in their restoration planning. One scientist who works on basin-wide modeling explained:

> It's really taking off. I think the reason that it hasn't been a big piece before is just that we didn't have the data that are in the models. So that's part of the story, right? We now have a lot of remote sensing and modeling techniques that just weren't there. We had no way. I remember when we started out on this case study in 2003 or so and everyone said, "It would be great if we had temperature models, how do we do that?" "Oh, that's too hard, forget it." But it's such a fundamental biological driver. It has to be there. And it's tied into climate change, obviously. So yes, now the challenge is how to manage all the data and interpret it correctly.

Temperature models like NorWeST StreamTemp, which bring together temperature and stream data from multiple spatial and temporal scales, are becoming critical to anticipating the future. Modeling efforts such as these, which predict stream temperature changes one hundred years into the future, are particularly useful for thinking about longer-term climate change across the region, allowing restorationists to identify critical areas for recovery. Yet for short-term and small-scale restoration work at the project-site scale, these models can also help anticipate potential impacts that may affect migrating populations upstream or downstream as temperatures change. Determining where to maintain or restore cool stream temperatures—often found in small pockets where ground water infiltrates—is important, and anticipatory, work. Identifying these cold-water refuges using remote sensing and stream temperature loggers is a now a priority for restorationists, and modeling techniques such as these can help restorationists hone in on locations that will make the biggest difference in terms of mitigating the effects of climate change. This ability, to tack between local and large scale and short and long term fills the gap in knowledge between decadal timescales and more finer-scale spatial patterns, helping to ensure the survival of a species from season to season as well as over the long term. These dynamics are becoming particularly important when considering climate change, and dealing with uncertainty and complexity in overlapping scales through the use of models is becoming a key strategy.

LIFE-CYCLE MODELS

Many restorationists believe that overlaying these stream temperature models with life-cycle models is the next step in understanding future habitat restoration needs in the Columbia River Basin. While not new to fisheries ecology, life-cycle models are becoming increasingly important in helping prioritize restoration efforts (Zabel, Cooney, & Jordan, 2013). According to NOAA Fisheries scientist Rich Zabel:

> A major advantage of life-cycle models is that they can translate changes in demographic rates (survival, capacity, or fecundity) in specific life stages into measures of population viability metrics (e.g., long-term abundance, productivity, or probability of

extinction), which are more relevant for population management. In addition, life-cycle models allow for the examination of impacts across several life stages in concert with other factors such as climate variability and change (Zabel, Cooney, & Jordan, 2013, p. 1).

Development of a robust model that covers all life stages of an anadromous fish, and then overlaying those life stages with climate change scenarios is a novel development that is still in progress. Life-cycle models in the Columbia River basin have become increasingly sophisticated, and now incorporate previously unknown parameters and objects of research such as "stochasticity, density dependence, and climate variability and change" (Zabel, Cooney, & Jordan, 2013, p. 1).

After the failure of the 2008 BiOp for the Columbia River hydropower system, which did not demonstrate to the courts that restoration efforts would actually have long-term effects on fish populations, NOAA Fisheries had to prove that it had thoroughly considered all influences in its modeling: incorporating climate effects, as well as habitat mitigation actions into life-cycle models in the Columbia River Basin became an important focus (Zabel, Cooney, & Jordan, 2013). The ultimate goal is that these models will make it possible to understand the effects of climate change on discreet populations. Because salmon often stray and the timing of their runs varies and overlaps, it has been challenging for scientists to classify salmon into discrete populations such as fall and spring runs. However, once populations are classified as separate, the hope is that life-cycle models can be used to "follow" individuals and populations of salmon as they migrate throughout the river. One modeler described the standardization process for grouping populations as a time-consuming, yet necessary, part of their work:

> A lot of the data management work is actually fairly tedious agreement on very specific protocols to figure out exactly what to call things and where the specific boundaries are [between populations]. But all that definitional work just allows you to roll the information up into units that you know are agreed to by everyone.

Again, this kind of data standardization is key to creating the knowledge infrastructure that will support adaptation because it will allow for more precise estimates of the effects of climate change on particular populations in particular stages of their life cycles. Once climate change models are layered on top of population models, a temporal resolution as small as a day will highlight the particular thermal niche that a fish is traveling through as it makes its way up the river. Many people expressed hope that these types of models will eventually be able to illustrate how temperature affects a population as it moves through a river system and identify potential problems like the warm waters that caused the massive die-off in 2015. Further, models like these can be used to predict when cold-water runoff will interact with particular kinds of habitats, allowing restorationists to identify priority locations for restoration where fish will need habitat most on their migration. For many restorationists, the "holy grail" of life-cycle modeling will be linking habitat monitoring and prioritization with life-cycle models, something that the court also called for after the 2008 BiOp (Zabel, Cooney, & Jordan, 2013). Further, by extending the timescale and incorporating climate change scenarios into life-cycle models, areas for restoration can be prioritized, and anticipating different futures for discreet populations of fish will be possible.

MODELS AS AN ANTICIPATORY STRATEGY

In order to overcome the increasing uncertainty that climate change introduces, modelers in the Columbia River Basin are developing increasingly sophisticated and scalable models. These two modeling efforts in particular—life-cycle modeling and stream temperature modeling—represent adaptive, anticipatory strategies that build in sociotechnical adaptability and produce results useful to the managers and institutions that rely on these results to make decisions.

Taming indeterminacy

By co-producing models with policy-makers and by choosing metrics that minimize and restrain uncertainty, modelers can "tame" some of the epistemic indeterminacy climate change introduces (van Hemert, 2013).

In the words of science studies scholar, Mieke van Hemert (2013) "adaptive approaches to biodiversity restoration require an adaptive epistemology, which implies tolerating a certain degree of epistemic indeterminacy" (p. 75). For example, one restorationist who was working in the field, described their effort to design a floodplain restoration project in this way:

> In spite of what the Independent Scientific Review Panel [of the NW Council] would like us to do, we can't monitor every single action. We just do not have enough people. That is really expensive and very time consuming. So, if we follow some models that have already been done and have been demonstrated to be effective, we are going to be pretty confident.

Yet it is important to remember that models are still based on field-collected data. Tim Beechie, a fish ecologist and restorationist who has worked in the basin for more than three decades, described the importance of field-collected data for parameter estimation in modeling:

> We've always gone to get data when we need it, and it can take a long time. But a lot of people either don't have the time, or maybe they don't get funding to go out and do data collection. The thing that I get nervous about is when we start to estimate things that we've never measured before to build models. If we know it's important, we'll go out and get the data if we need to.

Many restorationists described a similar tactic of drawing inference through a combination of pragmatic data collection, trained judgment, and modeling: a pragmatic move which—while not eliminating uncertainty—limits, or "tames" it so that decisions about how and what to restore can still be made. Models are important tools that facilitate this anticipatory work. As one restorationist pointed out, without modeling, anticipating different futures would be a far more "murky" process.

Dealing with Complex Scales

In a complex system, scale itself is important and being able to think about multiple scales at the same time is critical. As I have described, predictive modeling as a scientific practice requires tacking back and forth between regional-scale longer time-spans, and local-scale shorter time-spans. Conceptual tacking between temporal or spatial scales can occur between specific and abstract, local and global, and past, present, and future (Adams, Murphy, & Clarke, 2009). This movement between scales is captured in the way that scientific modeling is used to predict, and deal with, the complex nature of ecology and change (Taylor, 2005).

As another veteran restorationist and fish biologist put it: "If we really are in this for recovery of fish, let's look at it at the right spatial and temporal scale, and let's try our actions at that right temporal and spatial scale." Dealing with uncertainty is not only about creating finer detail models that can highlight local complexities at the site or on the river-reach scale. Instead, restorationists often rely on models to help understand longer-term diversity and larger-scale stochastic elements of the system. Hierarchical relationships can be built into models by including interactions across system scales. Multiple modeling tools can also be created to meet the needs of different users and the parameters of different scales so that restorationists can adapt their practices to meet uncertainty at a scale that is meaningful for their work. As climate change becomes more apparent, anticipating futures through more innovative modeling practices that use multiple scales is helping people deal with the complexity those multiple scales introduce.

While negotiation between scales is key to adaptive management and decision-making more generally, it also includes the changing practices that surface in everyday actions. One ecological modeler explained:

> We are trying to get a set of indicators that we can check back with and check up on and adapt in our planning based on how we see the trends moving through time. I don't do the policy side of things so I haven't seen how it happens. We repeat our work a lot . . . we did this really fast, quick, and dirty back-of-the-envelope version a couple years ago, and then a couple years later they go, "Okay now we have the tools to do this a little more dynamically so let's build a model and then validate it with fieldwork"—so I guess that's kind of just what we do.

In this way, restorationists anticipate the future while working in the present, accounting for and monitoring small-scale changes where they can, while at the same time bringing these observations to help build larger modeling efforts such as NorWeST StreamTemp. After all, the main impetus for creating the StreamTemp database was to bring together disparate and small-scale data collections to create a larger model that could inform decision-making across scales. As Isaak said, with a laugh: "As soon as we organized our first data set and created a website . . . everyone else said, 'wow, we want that too!' And so, we just said, 'sure, it will take us a while, but eventually we will work our way through it.'"

ANTICIPATION AND ADAPTIVE KNOWLEDGE INFRASTRUCTURES

This chapter describes how anticipation is another strategy that restorationists are using to reorient their work toward the future. In this case, it is illustrated by the ways in which they are taming indeterminacy and dealing with complex spatial and temporal scales through modeling. But, in doing so, restorationists are also employing scientific practices and creating knowledge infrastructures that facilitate the exploration of possible and anticipatory futures, while working from the present. An anticipatory strategy for science actively seeks to create flexible, resilient, and robust research and knowledge infrastructures, and it also tries to bring these futures into the present in both imagination as well as practice. As articulated by Adams et al. (2009), anticipatory scientific regimes occur when "sciences of *the actual* are displaced by *speculative forecast*." Ecological restorationists in the Columbia River Basin are carrying out "anticipation work" through their scientific practices by modeling spatial and temporal scales that take the long now into consideration (Ribes & Finholt, 2009).

Knowledge infrastructures can be standardized and scaled to better orient scientific work toward the long-term (Karasti et al., 2010). In this way, scientific work becomes allied with particular infrastructures that anticipate future spatial extents. As I described in chapter 5, setting standards is one way to acclimate scientific work to the future, but it is also a kind of "anticipation work" because it establishes what data will be comparable "across space and time" (Steinhardt & Jackson, 2015). In creating standards that orient toward the future, and by anticipating social and environmental change, more resilient scientific systems may be built. Many of the knowledge infrastructures that support ecological research and restoration in the Columbia River Basin are still evolving, and as described in the previous chapters, they are still under development or just being implemented. Now is therefore a good time to consider the adaptability of the infrastructures that support this work.

In terms of what it takes to design and build a sustainable knowledge infrastructure, then, we need to consider the long now of infrastructure design, as infrastructures are built to meet a particular goal, or vision of what the future should look like (Ribes & Finholt, 2009). By maintaining a common and interoperable data set across the basin—something the Pacific Northwest Aquatic Monitoring Partnership is working toward—the partnership is facilitating an adaptive knowledge infrastructure that will support the models needed to anticipate the future climate-changed and basin-wide environment. In other words, though the data used remains the same, the models themselves can change to meet new objects or conditions. This "kernel" of research infrastructure—or what Ribes (2014) identifies as "the core resources and services that an infrastructure makes available" to scientists, as well as the "work, techniques and technologies that" sustain those resources (p. 574)—allows both flexibility and extensibility of the endeavor into the future. Although this anticipatory work may not be flawless, it helps avoid difficult tasks, such as the "cleaning" work that Isaak's team was tasked with in creating the StreamTemp database.

Some restorationists are already orienting toward this "long now" of infrastructure. As one restoration consultant stated:

> We need to stop thinking about [research support] systems as a one-time capital expense, like: "I am going to go invest in this big database and it is going to take a few years and then be done with it." We've cautioned and tried to get people to think more long term. You really need to think about how you're going to have software that adapts and changes to your program—otherwise it becomes irrelevant or disused and just slowly dies a slow death.

So, we try to do whatever we can to try to caution people and get people to think about that.

By thinking in terms of a kernel of research infrastructure that can adapt to future needs, research programs can change to meet new technologies, objects of research, and environmental conditions (Ribes, 2014). Investing in databases that support modeling into the future is therefore one anticipatory strategy; but as I have described above, this is just one of the many ways that anticipation permeates the work that restorationists do.

ANTICIPATING FUTURES THROUGH MODELS AS METAPHORS

According to Stephanie Steinhardt and Steven Jackson (2015), anticipation work involves "practices that cultivate and channel expectations of the future, design pathways into those imaginations, and maintain those visions in the face of a dynamic world." These include practices that are able to connect individuals and cultures through this future "vision," as well as travel across spatial and temporal scales. *Practices for* the future and *imaginations about* the future come together in anticipation. Anticipatory practices help sensitize the field of restoration ecology to the future, and to emergent imaginations and ideas about the future that enable science-based decision-making in the present. This is a kind of "dual futurity" described by Weber (1946) in which humans are oriented toward the future and allow this future to guide their present actions. This reorientation constitutes a major epistemic and conceptual shift for a field like ecological restoration, which has been guided by restoring to the past for so long.

Following Steinhardt and Jackson (2015), "improvisation and adaptation are an integral part of anticipation work." Anticipation becomes a way of orienting oneself and one's work, but this must be supported by an anticipatory culture—through what Adams et al. (2009) call "regimes of anticipation," or cultures that "authorize speculative modes of engagement." In this vein, the scientific practice of modeling in ecological restoration is increasing in importance as climate change necessitates the development of new, future-oriented baselines and goals. According to Pickering (1995), "The goals of scientific practice are imaginatively transformed versions of the present. The future states of scientific culture at which practice aims are constructed from existing culture in a process of *modeling* (metaphor, analogy)" (p. 19). Pickering (1995) is referring to a broader notion of modeling in science than I am here. Still, through *predictive* modeling, anticipation becomes an important scientific practice—as well as a strategy for dealing with uncertainty and environmental change—by creating and enabling metaphors or, as I will describe in the final chapter: imaginaries.

By using models, restorationists are able to anticipate and "try out" different futures, empowering them to make choices about which restoration treatments to employ. They then monitor these actions to infer further and more accurately into the future. Predictive models make it possible to look to future states and to set future goals. This shift toward anticipating the future is necessary in order for restoration to be adaptive; and models are one strategy for anticipating and exploring these futures, especially in circumstances where data may be lacking. One restoration ecologist described this process:

> They'll run this model, using basically professional opinions, for conditions in each reach. They'll estimate how good they think it was historically and how good they think it is now, and they also might say, "What if we did some restoration? If we restore the riparian zone or if we remove these culverts, will that change things?" So, they use it in a smart way to figure out where the restoration opportunities are. . . . They use data where they can, but people don't measure things on every reach, and they don't measure every month on every reach.

Models create metaphors and analogies from which new practices can emerge. They make space for "what if?" questions. Pickering's (1995) description of models as "the link between existing culture and the future states that are the goals of scientific practice" is helpful here (p. 56). But he is careful not to be deterministic: "The link is not a causal or mechanical one: the choice of any particular model opens up an indefinite space of modeling vectors, of different goals" (p. 56). In this way, modeling as a scientific practice helps to bring about particular goals, sociotechnical imaginaries, or "future-natures" in the Columbia River Basin by anticipating the future, but not by determining it. Nevertheless, the choices that go into creating and using models, as well as the data that are fed into them, shape the way the river is restored into the future because they enable particular environmental and scientific imaginaries.

In this way, anticipation is also a normative force, orienting scientific work toward particular goals, some of which, as I have shown in previous chapters, are defined by legal institutions such as the Endangered Species Act. These institutions determine what kinds of scientific interventions are most appropriate and they authorize particular kinds of research through their requirements for particular types of knowledge. Modeling helps set restoration goals by facilitating the prioritization of restoration sites and ecological targets. In this way, anticipation facilitates a particular co-production of science and society. Through anticipatory norms and future imaginaries, knowledge infrastructures forge trajectories for scientific work. Anticipation "articulates" between the goals of institutions and scientific practices because anticipation predicts what is possible and creates openings for technoscientific interventions (Adams, Murphy, & Clarke, 2009).

Further, anticipation also has a moral quality to it. It embodies a moral injunction (Adams, Murphy, & Clarke, 2009) to look to the future and address the coming issue of climate change. This moral injunction to anticipate what will come is an important facet of the culture of ecological restoration, and it breaks with an epistemic cultural tradition of looking to the past and trying to minimize risk and uncertainty. As scientists orient themselves to a climate-changed future, anticipation is increasingly becoming an epistemic virtue, and it embodies a scientific culture whereby restoration science embraces the future, as well as the uncertainty it brings. For ecological restoration, which has been traditionally focused on looking to the past and using field-based methods, this shift to look to the future using modeling is potentially a big change for the epistemic community—but it also highlights how adaptive epistemologies can actually be.

The long timescale necessary to restore salmon to the Columbia River Basin presents an especially complex challenge to the epistemic community of restorationists who are working toward this goal, and throughout the coming decades, their practices will need to change to meet new challenges and incorporate new technologies. Restorationists acknowledge the intensification of increasingly uncertain conditions, as well as a lack of ready-made protocols to deal with them. Therefore, restoration practice is adapting, and strategies that anticipate the future are already being employed. As one ecological modeler pointed out: "As conditions in the river change . . . the methods that we can use are going to be different, and where we have to go to collect data is different . . . you have to adapt."

But it is also possible that the timescales that restoration projects employ at the planning level are far too short. A *very* long-term restoration plan may look fifty years into the future: but most plans only look to the next funding cycle, with project start and end dates only a few years apart. With the exception of a few of the region's long-term monitoring projects, or research focused on intensively monitored watershed sites, data will only be collected for a few years after a project's completion. This may prove to be one of the biggest blind spots in creating an adaptive restoration ecology. Time frames may need to be extended beyond what individuals usually conceive. Modeling is a key tool for addressing this shortcoming.

As a counterpoint to the short timescale of a discreet restoration *project*, the restoration *plans* of some of the Columbia River tribes represent longer timescales, and some ecologists would say the most realistic timescales, in the basin. One tribe's restoration plan, for instance, looks three hundred years into the future, and a restorationist who worked on developing the plan pointed out:

> An old-growth forest takes three hundred years to grow. This is something that I've struggled with personally. You go back to a restoration site that is ten years old, and it's exciting, but it's also really depressing to see those trees that you planted. You are never going to see them as old. My daughter might see them, but it's going to be a long time. [The plan] is a useful tool to say okay, these great problems are not going to be addressed in our lifetimes, but someone has to start doing it.

These kinds of anticipatory practices represent adaptive strategies that increasingly characterize and run throughout the restoration community in the Columbia River Basin, enabling restorationists to work intentionally toward alternative futures, or at the least to adapt to the one that is already arriving.

Ecological modeling practices tie together multiple temporal and spatial scales: past and future, local and global. Most importantly in terms of understanding how restorationists deal with climate change, modeling practices link present scientific practices to future goals (Pickering, 1995). The practice of modeling helps restorationists deal with complex spatial and temporal scales and allows them to "tame epistemic indeterminacy" (van Hemert, 2013). The practice of modeling also helps restorationists to envision future outcomes by creating metaphors of the future. These are all anticipatory practices and can be supported through the creation of adaptive knowledge infrastructures. Anticipation of the future follows the broader shift in restoration thinking described throughout this book. In the next chapter, I will build on the future-making concepts I have outlined—emergence, acclimation, and anticipation—to bring into focus the role of future sociotechnical and environmental imaginaries in science.

Conclusions

Environmental Imaginaries and River Futures

From where I was camping in the Teton Valley, I had to get an early start to make it all the way up to the crossroads they call Leadore, Idaho, in the Lemhi Valley. As I headed north and the sun rose, I drove along the Lemhi River as it winds its way through a broad valley at the base of the Bitterroot Mountains. Here, irrigated fields and ranches are cradled in the grass and shrubland of rolling foothills, which rise to meet a cloak of conifer forest. In the distance, the peaks of the Lemhi Range, which, at over 11,000 feet, are some of the tallest mountains in Idaho, showed faintly through the thick orange haze that was blanketing much of the state. Wildfires had been burning across the region throughout the summer, and last night, high winds and lightning strikes had ignited more fires in the valley. The smoke has become a regular feature of late summer for much of the Pacific Northwest. The ashen air offers beautiful orange sunsets, in addition to an overall apocalyptic feel. If the fires are ignited on wildlands where no private property is threatened, they are often left to burn—but this makes for a smoke-filled summer. Through the smoke, I could just make out the dots of black cattle grazing on the slopes below the tree line; above this, the craggy peaks with small patches of the late-summer remnants of snow.

I was on my way to meet Merrill Beyeler and Jeff DiLuccia, an unlikely pair who have been working together to restore salmon habitat to the Lemhi River and its tributary streams for well over a decade. DiLuccia is a fisheries biologist who has worked in the Lemhi valley for the Idaho Department of Fish and Game since 2004, when he was hired as the first "restoration coordinator" for the agency. Although DiLuccia has lived and worked here for a relatively long time, Merrill Beyeler and his family have been ranching in the valley for generations. In addition to being a rancher, Beyeler is a former schoolteacher and Idaho State Representative. He is clearly an educator at heart, possessing a rare, teacherly patience. Yet his passion and his livelihood come from ranching. He runs hundreds of cattle on these hills and, like many ranchers he believes that—when done right—ranching can benefit the landscape and provide a sustainable livelihood for rural people without harming fish habitat.

DiLuccia and Beyeler wanted to show me the changes that they had initiated in the riparian landscape in the valley. The Lemhi is home to one of the largest and arguably one of the most creative restoration projects I have visited in the basin. Early on, the project was conceived as a way to connect cool-water tributaries and create high-quality habitat for fish that used to spawn and rear in these high-elevation headwaters. The river had suffered from some of the same issues as the Tarboo Valley, which I visited with Peter Bahls and Sean Gallagher in chapter 3. The channel had been straightened and ditched in order to drain the floodplain and create hay fields, and overgrazing had degraded the stream banks in places. Some stretches of the stream had also been dewatered by irrigation to provide water for forage crops during the summer. The stream had been simplified and cut off from its floodplain. Lack of connection to groundwater had increased stream temperatures in a valley that is already feeling the effects of warming temperatures from climate change. As DiLuccia put it:

> Our focus is to provide better conditions in the face of climate change. The Lemhi was broken, and a lot of cooler water was not getting into the stream. We had lost a lot of the direct links to higher elevation habitats for trout and salmon.

DiLuccia is talking about the cold-water refugia that have been identified by stream temperature projects like NorWeST StreamTemp, refugia that are becoming very important for salmonids as climate change takes effect.

Together with the Lemhi Regional Land Trust and other ranchers and conservation groups in the valley, DiLuccia and Beyeler have restored over fourteen miles of what is predicted to be some of the most productive spawning and rearing habitat in the Columbia River Basin. The Lemhi is geologically unique in several ways, and reconnecting the river to its floodplain has some of the highest production potential of any stream in the Columbia River Basin (Bond et al., 2019). The river winds its way through a broad valley with very shallow bedrock, and the restoration project is restoring the meandering, braided stream that once existed. DiLuccia pointed out that lidar images have shown that "the river was all over that valley." In other words, it was a Stage 0 stream. In addition, high inputs of groundwater through a spring-fed system provide a stable temperature: warm in winter and cool in summer. High primary and secondary productivity make the Lemhi Valley an ideal place for salmon to rear, giving them a rich supply of food—similar to the habitat that Harlan Holmes recognized in Bear Valley, Idaho, back in 1927, when he wrote in his field diary: "the valley is level and comparatively flat . . . a meadow several miles wide in some places . . . it must be a wonderful nursery grounds for aquatic life upon which young salmon might feed" (Holmes, August 12, 1927). Indeed, Holmes even went up to the Lemhi in that same field trip: "there are a lot of nice fish there," he wrote in his diary (Holmes, August, 1927). But, since then, that hasn't always been the case.

The Lemhi restoration project is creative because it brings together some unlikely allies—ranchers, irrigators, and conservationists—and they have not only managed to work together, but they have been highly successful in doing so. By using a complex water transaction program—the Columbia Basin Water Transactions Program, facilitated by The Nature Conservancy and the Bonneville Power Administration—local landowners and irrigators have managed to ensure year-round flows in parts of the Lemhi that were previously dewatered by irrigation withdraws. But they have also experimented with creative ways to manage streamside grazing, allowing ranchers to maintain control of their core, river valley land and at the same time help improve the ecology of riparian areas. These grazing techniques keep cattle away from degraded steambanks at critical times and limit the number of cattle in floodplain areas. Through careful monitoring by the ranchers and ecologists, they have managed to increase water quality and numbers of spawning Chinook salmon. DiLuccia chalks this success up to the willingness of collaborators to seek creative solutions that allow habitat restoration and ranching to coexist: "The land owners and water users are key. Guys like Merrill get it. They can have both and can benefit from having both."

Over 5,000 feet and more than 800 miles from the mouth of the Columbia where it meets the Pacific Ocean, the Lemhi Valley, according to DiLuccia, is probably one of the highest elevation and most inland places in the world where anadromous fish migrate. Fish migrating to and from the Lemhi need to make it past countless obstacles, including dams. Yet it is these cool, headwater areas that are becoming ever more critical to salmon survival as climate change deepens.

In the Columbia River Basin, restorationists and ranchers alike are already experiencing climate change. They have been living through it for the last fifty years (Dalton, Mote, & Snover, 2013). The snowless winter, record long droughts, and high temperatures of 2015 were a window into one potential—and highly likely—future. The, increasingly frequent, smoky haze of wildfires that hangs throughout the region in late summer is yet another visceral reminder that climate change is occurring. These potential futures need to be anticipated and actions to adapt to them need to be purposefully considered and prepared for at all scales in the basin-wide effort to restore salmon. To help understand and prepare for these climate-altered futures, science is needed. But adaptability of scientific practice and knowledge infrastructures in enabling adaptation and change in socio-ecological systems has been underconceptualized. While gathering data and increasing knowledge will undoubtedly help facilitate decision-making, this book has argued that adaptation within epistemic communities themselves also requires attention. Instead of looking to science to "inform" policy independently, we also need to understand the complex role of the scientific effort itself in facilitating or hindering this process of adaptation, and especially the role of science in imagining and enabling different futures.

In the previous chapters, I brought together concepts from different disciplines in order to explore adaptation to environmental change in science. In doing so, I described three ways that I observed adaptation in the scientific practice of ecological restoration and the knowledge infrastructures that support these practices. I called these adaptations emergence, acclimation, and anticipation, and gave examples of how restorationists are employing them. The terms used—emergence, acclimation, and anticipation—were chosen because they bridge disciplines and can be applied to multiple scales of epistemic work. As concepts, they highlight the roles that knowledge production—and scientists themselves—play in adaptive change. It is important to point out that these are only some of the ways that scientists in the basin are adapting, and I am certain that there are many other strategies that are being used on a day-to-day basis as individuals cope with understanding change and deal with it in different epistemic communities throughout the world. This story from the field of restoration offers only one example of how an epistemic community can shift from looking to the past to looking to the future. But in different geographical and epistemic settings, there will no doubt be different tactics and strategies that appear. I offer these concepts only as a starting point to begin thinking about "adaptive epistemologies," both how to recognize them and how they could potentially be fostered in different scenarios where scientists are facing environmental or sociotechnical change and transitions.

Finally, I will offer three "conclusions," or, better yet, what, I hope, can serve as launching-off points for more work to be done on understanding "adaptive epistemologies." First, I hope that this exercise—in which I brought together concepts and theories from different fields—has demonstrated that inter- (and trans-) disciplinarity provides a productive opportunity and an important site for exploring complex environmental problems; and that using the tools that science and technology studies (STS) provides to look at what scientists are *doing* is especially helpful for making things visible in ways that can allow silenced voices to be heard. Second, this story of the field of ecological restoration's adaptation in the Columbia River Basin demonstrates that it is important to consider how collective goals for environmental management (whether manifested through policy or not) relate to science through an ongoing process of co-production of epistemic communities and legal regimes. Finally, in this concluding chapter, I want to reinvigorate the concept of "environmental imaginaries," which, I believe, is useful for thinking through how the ways in which we study and manage the environment can manifest in particular, material outcomes: different *river imaginaries* lead to different *river futures*.

MAKING VISIBLE

Science has always played a role in "reconfiguring 'the possible'" (Adams, Murphy, & Clarke, 2009). This is not the same as embracing a technological modernity in order to "solve" environmental and societal problems through "technological fixes." Instead, it is about recognizing that there are dynamics within science where different goals and intentions will facilitate different sociotechnical imaginaries. The example of co-production in setting monitoring standards in order to meet the requirements of the Endangered Species Act within the Columbia River Basin illustrates this. Some qualities of the ecosystem become elevated, more important, and more visible; while others, that are more complex or difficult to "see" through metrics—such as Stage 0 river valleys or macroinvertebrates—are more likely to remain uncounted and invisible.

The examples in the previous chapters demonstrate some of the ways in which ecological restorationists struggle with the uncertainty and the shifting goals, expectations, and physical disruptions that climate change introduces to their field. In naming the different strategies that restorationists are using to deal with environmental change, I want to make them visible. This ability to open up the "black box" of scientific practice is something that the conceptual and methodological tools of science and technology studies can provide. Only by being made visible can these systems be intentionally fostered by restorationists, managers, and policy-makers. These strategies can be seen in the scientific practices, the epistemic cultures, and the virtues that the restoration community holds. While these epistemic qualities can change due to shifts in individual practices, when talking about facilitating adaptive capacity, we must also consider the roles played by policy and management in enabling adaptive organizations and, especially, in supporting knowledge infrastructures. For example, by orienting their management plans, monitoring programs, and collaboration tools toward the future and, in doing so, preparing for climate change, restorationists are more likely to have the support that they will need to adapt and move forward in their work.

The same may be true for other scientific communities, such as those currently going through sociotechnical and environmental transitions, for example in fields such as renewable energy engineering, nanotechnology, or computing. Science studies scholar David Guston (2008) discusses an "anticipatory governance" for science, defined as "a broad-based capacity extended through society that can act on a variety of inputs to manage emerging knowledge-based technologies while such management is still possible" (vi.). This kind of governance, which can take place within scientific practice itself, can give voice to those "excluded from offering constructive visions of futures" (Guston, 2014, p. 218). While this is only a partial solution to creating more just futures, the tensions that are created when scientists and policy-makers reflect on their own roles in enabling futures can be a productive way to include more diverse futures, while also embracing both uncertainty and the political nature of science in a "self-conscious" way (Guston, 2014). This may also help us to avoid the trap of believing that the technological infrastructure in place—the hydropower dams that have transformed the Columbia into an "organic machine," for example—will necessarily determine what futures are possible. Although these dams have fundamentally rearranged our relationship with the nature of the Columbia Basin, these physical infrastructures are not only durable, but also, ultimately, fragile—requiring constant maintenance (Mitchell, 2014).

The things that restorationists choose to measure and standardize across the basin now will impact how nature is understood and valued in the future. This means that it is important that the future itself is anticipated and research infrastructures are developed that take the "long now" into account (Ribes & Finholt, 2009). One way to facilitate this is by intentionally designing flexible knowledge infrastructures, or building in technoscientific flexibility, so that these infrastructures can incorporate new objects of research into them as change occurs and new objects are made visible (Ribes & Polk, 2015). This means that we need to understand not only how and why new objects for study emerge (Rheinberger, 2010), but also how scientists can prepare for the inevitable arrival of climate change and the new scientific objects that will come with it. Making visible the strategies that restorationists are using to deal with these changes is one way to prepare for this arrival.

Studies such as this one, which incorporate historical and policy analysis, ethnographic methods, and participatory research, seek to make the things that scientists do visible. What scientists choose to measure and count, and how they classify those objects of research have material consequences (Bowker & Star, 2000). At one point in history, riparian areas themselves were a "residual category." They went unseen, uncared for, and were even destroyed (Fiege, 1999b; Langston, 2003). Restoration ecology is, in some ways, the story of these riparian areas becoming visible. Although classification of critical habitat and recovery planning through biological opinions is ultimately a matter for the courts, making things visible through scientific work is one of the ways that scientists influence these futures. Changing scientifically determined categories once they have been created can be a "cumbersome, bureaucratically fraught process" (Bowker & Star, 2000, p. 3), but this study demonstrates that classifications and measurements can change: that scientific work, and its supporting infrastructures are ultimately flexible. This is partly due to the ongoing work that is required to maintain these infrastructures (Bowker & Star, 2000), as well as the social and co-produced nature of scientific work itself. This is the important insight that science studies brings to the table: by looking at scientific practices, knowledge infrastructures, and the institutions and practices that shape them, we can see how science changes at each point. Carefully examining these sites of change is one way to observe the adaptation of epistemic work. This highlights the power of an interdisciplinary perspective in foregrounding what can otherwise often remain invisible.

What else needs to become visible in order to adapt to a climate-changed future? Further, what about the voices that have often been silenced by the institutions and organizations that exist today? These could be voices that have been intentionally suppressed because they didn't fit into the dominant narrative of a hydropower-altered-river driving prosperity: immigrant communities, those who have lost their natural resource livelihoods, or Indigenous peoples, for instance. Honoring tribal treaty rights and the rising recognition of tribes and their political power are examples of emerging and important shifts that are highlighting previously marginalized or silenced voices. Science's role in facilitating or hindering these more just futures should not be ignored. This includes honoring and making visible other ways of knowing and understanding the river.

CO-PRODUCING THE RIVER

The challenges that habitat restorationists in the Columbia River Basin face clarify that to sustainably manage socio-ecological systems, restorationists must be supported by adaptive knowledge infrastructures and institutions. The scientific institutions and organizations within the Columbia River Basin facilitate or hinder the production of particular kinds of knowledge, which, in turn, affects possibilities for adaptation to climate change. As I have shown, this is because these institutions rely on and mandate particular kinds of knowledge. As people come to understand the complexities involved in managing socio-ecological systems, increasing attention needs to be given to these institutions and organizations and their role in enabling or constraining adaptive capacity through co-production. This includes paying attention to the kinds of knowledge about the river that are being produced and valued. Attuning to co-production—through examining the relationships between law and science—can offer these conceptual insights.

Scientists not only provide information to these institutions, they are often tasked with translating their science to policy-makers and even advocating for particular scientific directions. As one restoration planner recounted restorationists' role in giving voice to science:

> I just try to always keep in mind those large external drivers to the systems, which when you are down in the trenches, you are not always thinking about. But it's about keeping these things in the forefront. In decision-making they take years to go from the conceptual to the actual. So, we say, "Hey, remember climate change? Remember invasive species and the food web?"

The interaction of restorationists and policy-makers goes beyond seeking funding and responding to calls for grant proposals. Instead, scientific practice involves constant communication—and collaboration—across disciplines and between scientists and institutions in the basin.

Scientists themselves are also advocating for changes in how scientific work is organized in the basin. One researcher and manager who oversaw salmon recovery in the Pacific Northwest for decades explained:

> You tend to have these groups that are built around certain science-policy outcomes and they become wedded together. In the late nineties that's what it was all about. There were major debates. We have improved and moved a long way past that. Now those camps are still set up, but there is more of an effort at collaboratively looking at these results. So, a couple of years ago the Federal Caucus hosted two workshops, one of which was on the impact of fire on salmon and watersheds and the other was on the issue of cold-water refuges in the Columbia Basin and what that meant. In the case of the fire one, we pulled together scientists from the tribes, from the Forest Service, and from a bunch of the different agencies and we had each done part of the research and came together with the findings. . . . So that idea of collaborative science is there.

This type of collaborative work is part of an evolving collective empiricism in the Columbia River Basin. More and more, people are embracing experimentation, improvisation, and interdisciplinarity so that novel strategies can emerge. While this book has not been a comprehensive study of virtues and ethics, trends emerged from the interviews to point to shifts in epistemic virtues moving toward a kind of collective empiricism. Daston and Galison, (2007) define "collective empiricism" as "the collaboration of investigators distributed over time and space in the study of natural phenomena too vast and various to be encompassed by a solitary thinker" (p. 27). In my research, I did see a kind of collective empiricism taking hold in the Columbia River Basin, which is indeed a place where the spatial extent and magnitude of the problem are far too large for individual restorationists to overcome on their own. Combining and standardizing data from across the basin into models and knowledge infrastructures that advance collective visions of the basin is one example of this trend.

Collective ways of working contrast with competitive modes (Gibbons et al., 1994). Large-scale scientific endeavors, especially, can lead to more collective ways of working because they rely on cooperation to succeed (Knorr-Cetina, 1999). These ways of working are also increasingly important in fields that grapple with high levels of uncertainty (Nowotny, Scott, & Gibbons, 2013). Facets of collective ways of working can be found in the restoration enterprise of the Columbia River Basin, as scientists and practitioners work toward the common goal of salmon habitat restoration. The mechanisms for implementing these collective modes include the practices discussed in this book, such as modeling, as a means for determining priorities and creating knowledge infrastructures that facilitate data-sharing. To be sure, I do not want to erase the many conflicts and often-discussed economic and even political competition between different actors within the Columbia River Basin. There are many and they are real. This is just to acknowledge a trend that many participants spoke about in interviews, applauded at conferences, and write into their organization's vision statements.

Across the field and throughout the basin, restorationists are fostering the emergence of alternative practices, pragmatically acclimating these practices, and anticipating possible futures. The field of restoration was born out of the idea that environmental repair was about looking to the past. Historic states of nature were used as baselines and goals for restoration projects. But over time, as climate change and other large-scale environmental shifts have become more pronounced, the discipline has had to transform to look forward and even to anticipate the future, becoming a more adaptive restoration science. This has been a crucial shift, but it has not come easily. Restorationists in the basin struggle with climate change and uncertainty; yet I have shown how—instead of remaining passive to these changes, which are being immediately felt—the epistemic community is actively adapting.

According to socio-ecological systems theory, adaptive institutions are flexible enough to deal with multiple and shifting objectives (Pahl-Wostl, 2009). Yet scientific institutions, as well as infrastructures, also contain their own "inertia" that can sometimes be difficult to shift, or even identify. Some restorationists in the basin worry that the institutional and organizational structure that exists is too rigid to deal with the novelty of climate change. However, a historical perspective helps highlight just how much these infrastructures, organizations, and institutions have already changed. Some of the restorationists that have been involved in the development of the field over the past two to three decades—from the time of ESA listing—have this arc of change in view. In the words of one the restorationists that I spoke to, whose career spanned this long term:

> If I knew fifteen years ago what I know now, we would have made a lot more progress. But I think that can be said of everybody that I work with. I didn't realize—and I don't think that any of us realized—fifteen years ago that we were basically redesigning an industry, and there was so much to learn. No, it's not *re*designing it *is* designing. We didn't have an industry! And if we had a chance to do it over we would do it a lot faster and we would do it a lot differently.

The task of salmon recovery is about much more than getting salmon back into the streams. It is also the task of developing the institutions and the knowledge infrastructures to support this recovery—and there was no road-map for the early restorationists to follow when they embarked on this effort.

Yet the recovery effort is also about collaboration, like that taking place in the Lemhi and in many other parts of the basin. This collaboration is, in many ways, very similar to the interdisciplinary adaptation occurring within the field of restoration itself, where people are more willing to "try anything" to see what solutions emerge. But this kind of interdisciplinary exercise does more than bring potential solutions to light through these collaborative interactions; it also surfaces different environmental imaginaries and brings them into conversation so that people can start to imagine, express, discuss, and collectively manage for different futures.

This interdisciplinary exercise echoes what Alfred North Whitehead (1927–1928) called "flights of speculative imagination," in which "the true method of discovery is like the flight of an aeroplane. It starts from the ground of a particular observation; it makes a flight in the thin air of imaginative generalization; and it again lands for renewed observation rendered acute by rational interpretation" (p. 5). Whitehead is best known as a mathematician, but later in his career he became a pathbreaking philosopher of science. His call for taking these "flights" highlights the creative leap that is necessary for the emergence of novelty; he is calling for interdisciplinarity. For Whitehead, science—indeed everything—is about creativity. The natural world is made of processes and is loaded with possibility. His work helps us recognize that certainty will not always serve us, and that "flights of speculative imagination," in which interdisciplinary and creativity can foster novel ideas, are integral to the emergence of possibility and action. Now might be a good time to embrace Whitehead's process-relational philosophy, as climate change stretches our ability to adapt and respond to environmental change and as scientists, natural resource managers, and communities call for new ways of addressing critical environmental problems.

ENVIRONMENTAL AND SOCIOTECHNICAL IMAGINARIES

While adaptations in scientific practice and knowledge infrastructures occur in response to environmental changes at the global scale, they are also shaped by how society collectively views and manages for the future through "sociotechnical imaginaries" (Jasanoff & Kim, 2015). Using a historical perspective to examine the development of ecological science has illustrated how societal values and "sociotechnical imaginaries" have influenced the ways in which science has been pursued and applied in the Columbia River Basin. Imaginaries are the "collectively held, institutionally stabilized, and publicly performed visions of desirable futures," and sociotechnical imaginaries are social formations enabled through science and technology (Jasanoff, 2015). Environmental management is realized through national scientific and natural resource policy programs that reflect a specific imaginary of the environment and how it should be managed to meet future national needs and interests. Much of the science that is produced at a particular time is created in response to a societal need for information, and these needs are based on a particular imaginary of what the future should be.

Holmes's early efforts to understand salmon and their habitats in the basin demonstrates how epistemic work is not only historically situated but is also conducted in response to societal needs. Yet in the case of environmental management like that in the Columbia River Basin, this imaginary is not only sociotechnical, but also environmental. I have shown that over the past several decades, a specific kind of restoration science has developed to meet institutional needs and societal goals. Between the time of Holmes's early travels and his later work at Bonneville Dam, sociotechnical and environmental imaginaries have shifted science in the basin from answering questions about where salmon are, what they do, and what they require to survive, to questions such as "how can salmon and dams coexist?" or "what kind of technology can we use to allow them to coexist?" It also demonstrates that scientific practice and the epistemic community that emerges to support that practice can also change. Therefore, the epistemic work and the normative goals and recommendations that are produced through ecological restoration will also shift when encountering environmental change. These changes are essential and confirm that science can be adaptive.

While the "sociotechnical imaginaries" literature hails from a science studies perspective, political ecologists have also developed analytical perspectives using "environmental imaginaries," or collective and situated ideas about nature (Peet & Watts, 2002). They describe how environmental management is realized through environmental imaginaries as scientific and natural resource policy programs come to reflect specific ideas about the future environment, as people work to manage it accordingly to meet future national or collective needs and interests (Nesbitt & Weiner, 2001). Environmental imaginaries highlight the ways in which different narratives and ways of knowing the environment can take hold, thereby facilitating different environmental futures. This can occur both through individual enactment of these imaginaries or by regulatory and management regimes that become institutionalized (Davis & Burke, 2011). It is important to be clear that imaginaries are not merely ideas in the mind; imaginaries are active in the world, reproducing environments in a particular way and having profound material and structural impacts (Mitchell, 2014).

Considering the role of both sociotechnical and environmental imaginaries brings focus to the ways that particular ways of knowing may be contested or adopted in management practice. The sociotechnical *and* environmental imaginary of the Columbia River Basin, for example, has changed dramatically over the last one hundred years. The river basin was transformed during the twentieth century into what White (1995) called an "organic machine," in which the Columbia and its tributaries were put to "use" through the creation of an industrial-scale hydropower complex. This development vision has had a profound effect on the kind of science being produced in the basin, as well as the baselines being used for that science. New imaginaries are also emerging, however, as environmental protection, tribal treaty rights, and environmental justice gain traction. The institutional and legal landscape as well as the epistemic work being done in the basin have brought different environmental baselines into and out of view over time.

"Environmental imaginaries" include discourses and practices concerning the relationship between nature, society, and development of natural resources, and are thus the site of contestation and formation of norms about the environment (Peet & Watts, 2002). These struggles are as much about how nature is known and understood as they are about how nature should be used (Peet & Watts, 2002). Imaginaries are also place-based and contextualized through environmental histories. When combined with the insights from sociotechnical imaginaries, the concept of environmental imaginaries provides a way to think about how particular futures become foreclosed through the enactment of scientific work. If sociotechnical imaginaries encode "visions of what is attainable through science and technology" (Jasanoff, 2015), similarly, environmental imaginaries set potential futures in play by enabling people to imagine what is possible. When combined in environmental management, our sociotechnical *and* environmental imaginaries therefore have material consequences for the futures that they further. While infrastructures are often viewed as the material and structural foundation on which scientific endeavors are built, imaginaries—despite their seemingly ephemeral nature—can also provide a strong basis from which knowledge is produced in particular ways through becoming embedded in the practices of individual scientists as well as the infrastructure, institutions, and organizations that support that particular knowledge.

Within a changed climate, future imaginaries for the field of ecological restoration are still emerging, and some new goals for restoring salmon to the Columbia River Basin are being discussed. One goal may be to simply aim for increasing ecological complexity itself. Another priority may be finding what will "move the dial" toward species recovery and focusing only on measures that lead to that. Yet another tactic may be to shift the priorities of some of the institutions in the basin, such as the Columbia River Treaty: adding the new goal of a functioning ecosystem to the current goals of hydropower and flood control (Cosens, 2010).

In order to recover salmon in the Columbia River Basin, environmental imaginaries for the river itself may also need to shift entirely. There is no doubt that there are multiple sociotechnical and environmental imaginaries operating within the basin, and it is highly unlikely that everyone will agree on a single imaginary and work toward it. In the words of one frustrated federal manager: "How much do we value salmon and steelhead and what are we willing to give up?" In a river basin that has been transformed in order to facilitate an industrial and economic boom powered by hydropower and the irrigation of the desert, there are too many interests at play to naively consider agreement possible. That is not the point of this exercise. Instead, it is to ask: what is the future river that we want to see, and in what ways does the science being conducted here create or hamper our effort to create this future river? Is it a future river that supports salmon and their ecosystems at "whatever the cost"? A future river that relies on technoscientific solutions to provide hydropower to the region while attempting to maintain fish populations? A future river ecosystem that survives the drastic environmental changes in store that temporarily became reality in the 2015 die-offs? Who gets to decide is a matter of politics and power; but as the Columbia River tribes have demonstrated through continual assertion of their treaty rights at multiple levels, environmental imaginaries have continuity, and when matched with political power can become powerful vectors for change.

A pragmatic adaptive management (Langston, 2003) of the basin would take future environmental and sociotechnical imaginaries into account, specifically considering what a future river will be. Environmental change will necessarily shift the way that science is being carried out, but the story of restoration in the Columbia River Basin shows that this is possible: the emergent quality of research itself requires that new categories, processes, and representations be incorporated as they arise (Ribes & Polk, 2015). This can be done through fostering an anticipatory practice in which we "think" and "live" toward the future (Adams et al., 2009). Adams et al. (2009) refer to this as a "politics of temporality," in which we have a moral responsibility to "secure the best possible futures" while recognizing that technoscientific futures "ratchet up" hopefulness and give possibility.

It is important to remember that once knowledge infrastructures, institutions, and organizations are created, they can also change. This is one of the findings of this research. Goals can be "tuned," as scientific practice unfolds (Pickering, 1995). As resistances such as climate change or societal shifts such as tribal treaty rights gain power, scientific goals must also be revised. By becoming aware of the dynamic nature of knowledge production, we open a new space for intentionality and adaptation within science itself.

In many instances in the Columbia River Basin there is little time to wait for answers about what to do or how to monitor or measure the effects of restoration: inaction is unacceptable. Therefore, restorationists are often forced into making decisions based on whatever information, experience, or intuition appears useful at the moment. If ecological restoration is to be successful in a climate-change-altered future, not only will restorationists themselves have to continue to adapt, but so will the knowledge infrastructures, institutions, and organizations that support them. Through a culture of anticipation, "the future arrives as already formed in the present, as if the emergency has already happened" (Adams et al., 2009). The "emergency" here is climate change, and people in the restoration community are adapting their cultures to anticipate it.

In this "transdisciplinary place," I will give the final word to Jennifer Molesworth, a restoration manager whose sentiment was one I heard echoed across the Columbia River Basin, from field sites to conferences, from city office towers to engineering firm headquarters:

> I think people often think that we are just going to go away and that this is a short term little shot of government largesse. But my mantra is that restoration is the future. The West has been discovered. Everything has been discovered. Now we have to rethink that, with our climate and all of our infrastructure. It is all going to be rebuilt. It all has to be made sustainable and

compatible with the natural environment. So, I think that, for young people starting out in their career, there is a really huge and bright future in trying to recover landscapes and watersheds and rivers and estuaries and the ocean. We have a lot of work to do. And none of it is impossible. We can fix everything. People really do know what is going on, I think. And I think that is an important thing, when we are looking at climate change and adaptation to climate change—that right now things may feel really negative, and mentally it is huge. I think there is a little bit of pushback, but we have no choice.

NOTES

INTRODUCTION

1 While most forms of restoration look to the past for reference ecosystems, river restoration is not only backward facing. As I will discuss in later chapters, many restoration tactics employ the processes of dynamism and change of rivers themselves to transform them. Nevertheless, as I will describe, river restorationists still look to the past to establish baselines and locate historic river channels, as well as use reference sites that are historic in nature.

CHAPTER 1: A SCIENCE FOR THE COLUMBIA RIVER BASIN

1 The Harlan Holmes Papers, from which these excerpts are drawn, are located in University of Washington Libraries Special Collections, Harlan Holmes Papers, Acc. 2614-001.
2 Some further reading on this topic includes M. Fiege, *Irrigated Eden: The making of an agricultural landscape in the American West* (1999); B. Cosens (Ed.), *The Columbia River treaty revisited: Transboundary river governance in the face of uncertainty* (1999); J. E. Taylor III, *Making salmon: An environmental history of the Northwest fisheries crisis* (1999); R. White, *The organic machine: The remaking of the Columbia River* (1995); and K. Barber, *Death of Celilo Falls* (2005).

CHAPTER 2: RIVER RESTORATION
IN THE COLUMBIA RIVER BASIN

1 Hatchery supplementation, which is a form of salmon population "recovery" in some areas, is also sometimes referred to as "restoration;" but throughout this study, I focus on ecological—or habitat—restoration, and do not include hatchery supplementation in my definition of restoration.
2 Treaties with the US government include: Treaty with the Yakama, 1855; Treaty with the Tribes of Middle Oregon, 1885; Treaty with the Walla Walla, Cayuse, and

others, 1855; Treaty with the Nez Perces, 1855; among many others in the Columbia River Basin.

CHAPTER 3: THE WORK OF RESTORATION IN A CHANGING CLIMATE

1. The Tarboo Valley River restoration, which is referenced in the story throughout this chapter, is not within the Columbia River Basin. It is a coastal watershed on the other side of the Cascade Range. Regardless, I have decided to include it as a narrative because the restoration project is considered one of the largest and most extensive habitat restoration projects in the region. It is also relevant to the restoration taking place in the Columbia River Basin because the ecological, historical, and social dynamics are very similar. Many people working within the Columbia River Basin are aware of the Tarboo Valley project.
2. The following narrative includes quotes from Gallagher, and while Bahls couldn't make the tour, he was interviewed later.

REFERENCES

Abatzoglou, J. T. (2013). Development of gridded surface meteorological data for ecological applications and modelling. *International Journal of Climatology, 33*, 121–131.

Adams, C. C. (1913). *Guide to the study of animal ecology*. New York: Macmillan.

Adams, V., Murphy, M., & Clarke, A. E. (2009). Anticipation: Technoscience, life, affect, temporality. *Subjectivity, 28*(1), 246–265. https://doi.org/10.1057/sub.2009.18

Alagona, P. S. (2013). *After the grizzly: Endangered species and the politics of place in California*. Berkeley: University of California Press.

Allen, C. (2003). Replacing salmon: Columbia River Indian fishing rights and the geography of fisheries mitigation. *Oregon Historical Quarterly, 104*(2), 196–227.

Allison, S. K. (2012). *Ecological restoration and environmental change: Renewing damaged ecosystems*. Hoboken, NJ: Taylor & Francis.

Anderson, B. (2006). *Imagined communities: Reflections on the origin and spread of nationalism*. London: Verso.

Ankersen, T. T., & Regan, K. E. (2010). Shifting baselines and backsliding benchmarks: The need for the National Environmental Legacy Act to address the ecologies of restoration, resilience, and reconciliation. In A. C. Flournoy & D. M. Driesen (Eds.), *Beyond environmental law: Policy proposals for a better environmental future* (pp. 53–80). Cambridge, UK: Cambridge University Press.

Aronson, J., & Alexander, S. (2013). Ecosystem restoration is now a global priority: Time to roll up our sleeves; News report from CBDCOP11. *Restoration Ecology, 21*(3), 293–296. https://doi.org/10.1111/rec.12011

Atkinson, C. (1988). The Montlake Laboratory of the Bureau of Commercial Fisheries and its biological research, 1931–81. *Marine Fisheries Review, 50*(4), 97–110.

Baker, S., & Eckerberg, K. (2013). A policy analysis perspective on ecological restoration. *Ecology and Society, 18*(2). https://doi.org/10.5751/ES-05476-180217

Barber, K. (2005). *Death of Celilo Falls*. Seattle: University of Washington Press.

Bash, J., & Ryan, C. M. (2002). Stream restoration and enhancement projects: Is anyone monitoring? *Environmental Management, 29*, 877–885. https://doi.org/10.1007/s00267-001-0066-3

Battin, J., Wiley, M. W., Ruckelshaus, M. H., Palmer, R. N., Korb, E., Bartz, K. K., & Imaki, H. (2007). Projected impacts of climate change on salmon habitat restoration. *Proceedings of the National Academy of Sciences, 104*(16), 6720–6725.

Beechie, T. J., & Bolton, S. (1999). An approach to restoring salmonid habitat-forming processes in Pacific Northwest watersheds. *Fisheries, 24*(4), 6–15.

Beechie, T. J., Imaki, H., Greene, J., Wade, A., Wu, H., Pess, G., . . . & Mantua, N. (2013). Restoring salmon habitat for a changing climate. *River Research and Applications, 29*(8), 939–960. https://doi.org/10.1002/rra.2590

Beechie, T. J., Sear, D. A., Olden, J. D., Pess, G. R., Buffington, J. M., Moir, H., . . . & Pollock, M. M. (2010). Process-based principles for restoring river ecosystems. *BioScience, 60*(3), 209–222.

Bennett, S., Pess, G., Bouwes, N., Roni, P., Bilby, R., Gallagher, S., . . . & Greene, C. (2016). Progress and challenges of testing the effectiveness of stream restoration in the Pacific Northwest using intensively monitored watersheds. *Fisheries, 41*, 92–103. https://doi.org/10.1080/03632415.2015.1127805

Bernazzani, P., Bradley, B. A., & Opperman, J. J. (2012). Integrating climate change into habitat conservation plans under the U.S. Endangered Species Act. *Environmental Management, 49*(6), 1103–1114. https://doi.org/10.1007/s00267-012-9853-2

Bernhardt, E. S., & Palmer, M. A. (2011). River restoration: The fuzzy logic of repairing reaches to reverse watershed scale degradation. *Ecological Applications, 21*, 1926–1931.

Bernhardt, E. S., Palmer, M. A., Allan, J. D., Alexander, G., Barnas, K., Brooks, S., . . . & Sudduth, E. (2005). Synthesizing U.S. river restoration efforts. *Science, 308*(5722), 636–637. https://doi.org/10.1126/science.1109769

Blumm, M. C., & Paulsen, A. (2012). The role of the judge in Endangered Species Act implementation: District Judge James Redden and the Columbia Basin salmon saga (2012). *Stanford Environmental Law Journal, 32*(87), 2013; Lewis & Clark Law School Legal Studies Research Paper No. 2012-12. http://doi.org/10.2139/ssrn.2051638

Blunden, J., & Arndt, D. S. (2015). State of the climate in 2015. *Bulletin of the American Meteorological Society, 97*(8), 300.

Bocking, S. (2004). *Nature's experts: Science, politics, and the environment.* New Brunswick, NJ: Rutgers University Press.

Bond, M. H., Nodine, T. G., Beechie, T. J., & Zabel, R. W. (2019). Estimating the benefits of widespread reconnection for Columbia River Chinook salmon. *Canadian Journal of Fisheries and Aquatic Sciences, 76*, 1212–1226. http://dx.doi.org/10.1139/cjfas-2018-0108

Bond, N. A., Cronin, M. F., Freeland, H., & Mantua, N. (2015). Causes and impacts of the 2014 warm anomaly in the NE Pacific. *Geophysical Research Letters, 42*(9), 3414–3420.

Bouwes, N., Weber, N., Jordan, C. E., Saunders, W. C., Tattam, I. A., Volk, C., . . . & Pollock, M. M. (2016). Ecosystem experiment reveals benefits of natural and simulated beaver dams to a threatened population of steelhead (*Oncorhynchus mykiss*). *Scientific Reports, 6*, 28581. https://doi.org/10.1038/srep28581

Bowker, G., Baker, K., Millerand, F., & Ribes, D. (2010). Toward information infrastructure studies: Ways of knowing in a networked environment. In J. Hunsinger, L. Klastrup, & M. Allen (Eds.), *International handbook of internet research* (pp. 97–117). Heidelberg: Springer.

Bowker, G., & Star, S. L. (2000). *Sorting things out: Classification and its consequences.* Cambridge, MA: MIT Press.

BPA. (2019, January). *Fact sheet: BPA invests in fish and wildlife.* www.bpa.gov/news/pubs/FactSheets/fs-201901-BPA-invests-in-fish-and-wildlife.pdf

Bradshaw, A. D. (1987). Restoration: An acid test for ecology. In W. R. Jordan III, M. E. Gilpin, & J. D. Aber (Eds.), *Restoration ecology: A synthetic approach to ecological research.* Cambridge, UK: Cambridge University Press.

Braverman, I. (2018). *Coral whisperers: Scientists on the brink.* Oakland: University of California Press.

Bromley, D. W. (2008). Volitional pragmatism. *Ecological Economics, 68*(1–2), 1–13. https://doi.org/10.1016/j.ecolecon.2008.08.012

Butler, D. H., Koivisto, S., Brumfeld, V., & Shahack-Gross, R. (2019). Early evidence for northern salmonid fisheries discovered using novel mineral proxies. *Scientific Reports, 9*, 147.

Cabin, R. J. (2007a). Science and restoration under a big, demon haunted tent: Reply to Giardina et al. (2007). *Restoration Ecology, 15*(3), 3777–3381.

Cabin, R. J. (2007b). Science-driven restoration: A square grid on a round earth? *Restoration Ecology, 15*(1), 1–7. https://doi.org/10.1111/j.1526-100X.2006.00183.x

Cabin, R. J. (2011). *Intelligent tinkering: Bridging the gap between science and practice.* Washington, DC: Island Press.

Cash, D. W., Clark, W. C., Alcock, F., Dickson, N. M., Eckley, N., Guston, D. H., . . . & Mitchell, R. B. (2003). Knowledge systems for sustainable development. *Proceedings of the National Academy of Sciences, 100*(14), 8086–8091. https://doi.org/10.1073/pnas.1231332100

Chaffin, B. C., & Gunderson, L. H. (2016). Emergence, institutionalization and renewal: Rhythms of adaptive governance in complex social-ecological systems. *Journal of Environmental Management, 165*, 81–87. https://doi.org/10.1016/j.jenvman.2015.09.003

Charmaz, K. (2005). Grounded theory in the 21st century: A qualitative method for advancing social justice research. In N. K. Denzin & Y. S. Lincoln (Eds.), *Handbook of qualitative research* (3rd ed., pp. 507–536). Thousand Oaks, CA: Sage Publications.

Choi, Y. D. (2004). Theories for ecological restoration in changing environment: Toward "futuristic" restoration. *Ecological Research, 19*(1), 75–81. https://doi.org/10.1111/j.1440-1703.2003.00594.x

Choi, Y. D. (2007). Restoration ecology to the future: A call for new paradigm. *Restoration Ecology, 15*(2), 351–353. https://doi.org/10.1111/j.1526-100X.2007.00224.x

Clarke, A. (2005). *Situational analysis: Grounded theory after the postmodern turn.* Thousand Oaks, CA: Sage Publications.

Clewell, A. (2009). Intent of ecological restoration, its circumscription, and its standards. *Ecological Restoration, 27*(1), 5–7. https://doi.org/10.3368/er.27.1.5

Cluer, B., & Thorne, C. (2014). A stream evolution model integrating habitat and ecosystem benefits. *River Research and Applications, 30*(2), 135–154. https://doi.org/10.1002/rra.2631

Committee on Protection and Management of Pacific Northwest Anadromous Salmonids. (1996). *Upstream: Salmon and society in the Pacific Northwest*. Washington, DC: National Academy Press.

Conroy, M. J., Runge, M. C., Nichols, J. D., Stodola, K. W., & Cooper, R. J. (2011). Conservation in the face of climate change: The roles of alternative models, monitoring, and adaptation in confronting and reducing uncertainty. *Biological Conservation, 144*(4), 1204–1213. https://doi.org/10.1016/j.biocon.2010.10.019

Corlett, R. T. (2016). Restoration, reintroduction, and rewilding in a changing world. *Trends in Ecology & Evolution, 31*(6), 453–462. https://doi.org/10.1016/j.tree.2016.02.017

Correll, Lt. Col. L. W. (1952, October 29). Letter from L. W. Correll to Leo. L. Laythe. Box 2, US Army Engineers Corps, Correspondence. Harlan Holmes Papers, Acc. 2614-001, Special Collections, University of Washington Libraries.

Cosens, B. (2008). Resolving conflict in non-ideal, complex systems: Solutions for the law-science breakdown in environmental and natural resource law. *Natural Resources Law Journal, 48*, 257.

Cosens, B. (2010). Transboundary river governance in the face of uncertainty: Resilience theory and the Columbia River Treaty. *Journal of Land Resources and Environmental Law, 30*, 229–265.

Cosens, B. (2012). Changes in empowerment: Rising voices in Columbia Basin resource management. In B. Cosens (Ed.), *The Columbia River Treaty revisited* (pp. 61–68). Corvallis: Oregon State University Press.

Cosens, B. (2013). Legitimacy, adaptation, and resilience in ecosystem management. *Ecology and Society, 18*(1). https://doi.org/10.5751/ES-05093-180103

Cosens, B., Craig, R., Hirsch, S. L., Arnold, C. A., Benson, M., DeCaro, D., . . . & Schlager, E. (2017). The role of law in adaptive governance. *Ecology and Society, 22*(1). https://doi.org/10.5751/ES-08731-220130

Cosens, B., Gunderson, L., Allen, C., & Benson, M. (2014). Identifying legal, ecological and governance obstacles, and opportunities for adapting to climate change. *Sustainability, 6*(4), 2338–2356. https://doi.org/10.3390/su6042338

Cosens, B., & Williams, M. (2012). Resilience and water governance: Adaptive governance in the Columbia River Basin. *Ecology and Society, 17*(4). https://doi.org/10.5751/ES-04986-170403

Craig, R. K., & Ruhl, J. B. (2014). Designing administrative law for adaptive management. *Vanderbilt Law Review, 67*, 1–87; Vanderbilt Public Law Research Paper No. 13-6; University of Utah College of Law Research Paper No. 18. http://doi.org/10.2139/ssrn.2222009

CRITFC. (1995). *Spirit of the salmon: Wy-Kan-Ush-Mi Wa-Kish-Wit.* https://plan.critfc.org

Cronin, K. (2008). *Transdisciplinary research (TDR) and sustainability* (Overview report prepared for the Ministry of Research, Science and Technology). https://learningforsustainability.net/pubs/Transdisciplinary_Research_and_Sustainability.pdf

Cronon, W. (1983). *Changes in the land: Indians, colonists and the ecology of New England.* New York: Hill and Wang.

Cronon, W. (1992). *Nature's metropolis: Chicago and the Great West.* New York: W. W. Norton.

Crozier, L. (2016). *Impacts of climate change on salmon of the Pacific Northwest: A review of the scientific literature published in 2015.* Seattle: NOAA Northwest Fisheries Science Center.

Dalton, M., M., Mote, P., W., & Snover, A. K. (2013). *Climate change in the Northwest: Implications for our landscapes, waters, and communities.* Washington, DC: Island Press.

Daston, L., & Galison, P. (2007). *Objectivity.* New York: Zone Books.

Dauble, D. D., Hanrahan, T. P., Geist, D. R., & Parsley, M. J. (2003). Impacts of the Columbia River hydroelectric system on main-stem habitats of fall Chinook salmon. *North American Journal of Fisheries Management, 23*(3). https://doi.org/10.1577/M02-013

Davies, N. S., & Gibling, M. R. (2011). Evolution of fixed-channel alluvial plains in response to Carboniferous vegetation. *Nature Geoscience, 4*(9), 629–633. https://doi.org/10.1038/ngeo1237

Davis, D. K., & Burke, E. (2011). *Environmental imaginaries of the Middle East and North Africa.* Athens: University of Ohio Press.

Doremus, H. (2001). Adaptive management, the Endangered Species Act, and the institutional challenges of "new age" environmental protection. *Washburn Law Journal, 41,* 50–89.

Doremus, H., & Tarlock, A. D. (2005). Science, judgment, and controversy in natural resource regulation. *Public Land and Resources Law Review, 26,* 1–51; UC Davis Legal Studies Research Paper No. 50. https://ssrn.com/abstract=788045

Douglas, H. (2000). Inductive risk and values in science. *Philosophy of Science, 67*(4), 559–579.

Duncan, D. H., & Wintle, B. A. (2008). Towards adaptive management of native vegetation in regional landscapes. In C. Pettit, I. Bishop, W. Cartwright, D. Duncan, K. Lowell, & D. Pullar (Eds.), *Landscape analysis and visualization: Spatial models for natural resource management and planning* (pp. 159–182). Berlin: Springer.

Dunwiddie, P. W., Hall, S. A., Ingraham, M. W., Bakker, J. D., Nelson, K. S., Fuller, R., & Gray, E. (2009). Rethinking conservation practice in light of climate change. *Ecological Restoration, 27*(3), 320–329. https://doi.org/10.3368/er.27.3.320

Eden, S., & Tunstall, S. (2006). Ecological versus social restoration? How urban river restoration challenges but also fails to challenge the science-policy nexus in the United Kingdom. *Environment and Planning C: Government and Policy, 24*(5), 661–680.

Edwards, P. (2003). Infrastructure and modernity: Force, time, and social organization in the history of sociotechnical systems. In T. J. Misa, P. Brey, & A. Feenberg (Eds.), *Modernity and technology* (pp. 185–225). Cambridge, MA: MIT Press.

Edwards, P. (2010). *A vast machine: Computer models, climate data, and the politics of global warming.* Cambridge, MA: MIT Press.

Edwards, P., Jackson, S. J., Chalmers, M. K., Bowker, G., Borgman, C. L., Ribes, D., . . . & Calvert, S. (2013). *Knowledge infrastructures: Intellectual frameworks and research challenges.* Ann Arbor: Deep Blue. http://hdl.handle.net/2027.42/97552

Egan, D. (1990). Historic initiatives in ecological restoration. *Restoration & Management Notes, 8*(2), 83–89.

Ehrenfeld, J. G. (2000). Defining the limits of restoration: The need for realistic goals. *Restoration Ecology, 8*(1), 2–9.

Evers, C., Wardropper, C. B., Branoff, B., Granek, E., Hirsch, S. L., Link, T., . . . & Wilson, C. (2018). The ecosystem services and biodiversity of novel ecosystems: A literature review. *Global Ecology and Conservation, 13,* 1–11.

Fiege, M. (1999a). Creating a hybrid landscape: Irrigated agriculture in Idaho. In D. D. Goble & P. W. Hirt (Eds.), *Northwest lands, Northwest peoples: Readings in environmental history* (pp. 362–388). Seattle: University of Washington Press.

Fiege, M. (1999b). *Irrigated Eden: The making of an agricultural landscape in the American West.* Seattle: University of Washington Press.

Folke, C., Hahn, T., Olsson, P., & Norberg, J. (2005). Adaptive governance of social-ecological systems. *Annual Review of Environment and Resources, 30*(1), 441–473. https://doi.org/10.1146/annurev.energy.30.050504.144511

Foucault, M. (1972). *The archaeology of knowledge: And the discourse on language.* New York: Pantheon Books.

Freeman, S. (2018). *Saving Tarboo Creek: One family's quest to heal the land.* Portland, OR: Timber Press.

Funtowicz, S. O., & Ravetz, J. R. (1993). Science for the post-normal age. *Futures, 25*(7), 739–755. https://doi.org/10.1016/0016-3287(93)90022-L

Garmestani, A. S., Allen, C. R., & Benson, M. H. (2013). Can law foster social-ecological resilience? *Ecology and Society, 18*(2). https://doi.org/10.5751/ES-05927-180237

Geissler, P. W., & Kelly, A. H. (2016). A home for science: The life and times of tropical and polar field stations. *Social Studies of Science, 46*(6), 797–808. https://doi.org/10.1177/0306312716680767

Gewin, V. (2015). North Pacific "blob" stirs up fisheries management. *Nature, 524,* 396. https://doi.org/10.1038/nature.2015.18218

Giardina, C. P. (2007). Science driven restoration: A candle in a demon haunted world. *Restoration Ecology, 15*(2), 171–176.

Gibbons, M., Limoges, C., Nowotny, H., Schwartzman, S., Scott, P., & Trow, M. (1994). *The new production of knowledge: The dynamics of science and research in contemporary societies.* London: Sage Publications.

Glacken, C. J. (1990). *Traces on the Rhodian shore: Nature and culture in Western thought from ancient times to the end of the Eighteenth Century* (5th ed.). Berkeley: University of California Press. (Originally published 1967)

Goble, D. (1999). Salmon in the Columbia Basin: From abundance to extinction. In D. D. Goble and P. W. Hirt (Eds.), *Northwest lands, Northwest peoples: Readings in environmental history* (pp. 229–263). Seattle: University of Washington Press.

Gosnell, H., Chaffin, B. C., Ruhl, J. B., Arnold, C. A. (Tony), Craig, R. K., Benson, M. H., & Devenish, A. (2017). Transforming (perceived) rigidity in environmental law through adaptive governance: A case of Endangered Species Act implementation. *Ecology and Society, 22*(4). https://doi.org/10.5751/ES-09887-220442

Gross, M. (2010). *Ignorance and surprise: Science, society, and ecological design*. Cambridge, MA: MIT Press.

Gunderson, L. (2013). How the Endangered Species Act promotes unintelligent, misplaced tinkering. *Ecology and Society, 18*(1). https://doi.org/10.5751/ES-05601-180112

Gunderson, L. H., & Holling, C. S. (2002). *Panarchy: Understanding transformations in systems of humans and nature*. Washington, DC: Island Press.

Gunderson, L. H., & Light, S. (2006). Adaptive management and adaptive governance in the Everglades ecosystem. *Policy Sciences, 39*, 323–334.

Guston, D. H. (2008). Preface. In E. Fisher, C. Selin, & J.M. Wetmore (Eds.), *The yearbook of nanotechnology in society: Presenting futures* (vol. 1, pp. v–viii). New York: Springer.

Guston, D. H. (2014). Understanding "anticipatory governance." *Social Studies of Science, 44*(2), 218–242.

Haas, P. (1990). *Saving the Mediterranean: The politics of international environmental cooperation*. New York: Columbia University Press.

Haas, P. (1992). Introduction: Epistemic communities and international policy coordination. *International Organization, 46*(1), 1–35

Haas, P. (2008). Epistemic communities. In D. Bodansky, J. Brunnee, & E. Hey (Eds.), *The Oxford handbook of international environmental law* (pp. 792–807). Oxford: Oxford University Press.

Hacking, I. (1992). The self-vindication of the laboratory sciences. In A. Pickering (Ed.), *Science as practice and culture* (pp. 29–64). Chicago: University of Chicago Press.

Haider, L. J., Hentati-Sundberg, J., Giusti, M., Goodness, J., Hamann, M., Masterson, V. A., . . . & Sinare, H. (2018). The undisciplinary journey: Early-career perspectives in sustainability science. *Sustainability Science, 13*(1), 191–204. https://doi.org/10.1007/s11625-017-0445-1

Hall, M. (2005). *Earth repair: A transatlantic history of environmental restoration*. Charlottesville: University of Virginia Press.

Hall, S. (1997). The work of representation. In S. Hall (Ed.), *Representation: Cultural representations and signifying practices* (pp. 13–69). Thousand Oaks, CA: Sage Publications.

Hansen, L., Hoffman, J., Drews, C., & Mielbrecht, E. (2010). Designing climate-smart conservation: Guidance and case studies. *Conservation Biology*, *24*(1), 63–69. https://doi.org/10.1111/j.1523-1739.2009.01404.x

Haraway, D. (1991). Situated knowledges: The science question in feminism and the privilege of partial perspective. In *Simians, cyborgs, and women: The reinvention of nature*. New York: Routledge.

Harris, J. A., Hobbs, R. J., Higgs, E., & Aronson, J. (2006). Ecological restoration and global climate change. *Restoration Ecology*, *14*(2), 170–176. https://doi.org/10.1111/j.1526-100X.2006.00136.x

Heller, N. E., & Zavaleta, E. S. (2009). Biodiversity management in the face of climate change: A review of 22 years of recommendations. *Biological Conservation*, *142*(1), 14–32. https://doi.org/10.1016/j.biocon.2008.10.006

Herrick, C., & Sarewitz, D. (2000). Ex post evaluation: A more effective role for scientific assessments in environmental policy. *Science, Technology, & Human Values*, *25*(3), 309–331.

Heter, E. (1950). Transplanting beavers by airplane and parachute. *Journal of Wildlife Management*, *14*(2), 143–147.

Hiers, J. K., Mitchell, R. J., Barnett, A., Walters, J. R., Mack, M., Williams, B., & Sutter, R. (2012). The dynamic reference concept: Measuring restoration success in a rapidly changing no-analogue future. *Ecological Restoration*, *30*(1), 27–36. https://doi.org/10.3368/er.30.1.27

Higgs, E. (1994). Expanding the scope of restoration ecology. *Restoration Ecology*, *2*, 137–146.

Higgs, E. (2003). *Nature by design: People, natural process, and ecological restoration*. Cambridge, MA: MIT Press.

Higgs, E. (2005). The two-culture problem: Ecological restoration and the integration of knowledge. *Restoration Ecology*, *13*, 159–164.

Hirsch, S. (2019). Anticipatory practices: Shifting baselines and environmental imaginaries of ecological restoration in the Columbia River Basin. *Environment and Planning E: Nature and Space*, *3*(1), 40–57.

Hirsch, S., & Long, J. A. (2018). Non-epistemic values in adaptive management: Framing possibilities in the legal context of endangered Columbia River salmon. *Environmental Values*, *27*, 467–488.

Hirsch, S., & Long, J. A. (2020). Adaptive epistemologies: Conceptualizing adaptation to climate change in environmental science. *Science, Technology, & Human Values*. https://doi.org/10.1177/0162243919898517

Hirt, P. W., & Sowards, A. M. (2012). The past and future of the Columbia River. In *The Columbia River Treaty revisited: Transboundary river governance in the face of uncertainty*. Corvallis: Oregon State University Press.

Hobbs, R. J., Arico, S., Aronson, J., Baron, J. S., Bridgewater, P., Cramer, V. A., . . . & Zobel, M. (2006). Novel ecosystems: Theoretical and management aspects of the new ecological world order. *Global Ecology and Biogeography*, *15*(1), 1–7. https://doi.org/10.1111/j.1466-822X.2006.00212.x

Hobbs, R. J., & Cramer, V. A. (2008). Restoration ecology: Interventionist approaches for restoring and maintaining ecosystem function in the face of rapid environmental change. *Annual Review of Environment and Resources, 33*(1), 39–61. https://doi.org/10.1146/annurev.environ.33.020107.113631

Hobbs, R. J., Higgs, E. S., & Hall, C. M. (2013). Introduction: Why novel ecosystems? In n R. J. Hobbs, E. S. Higgs, & C. M. Hall (Eds.), *Novel ecosystems* (pp. 3–8). Chichester, UK: John Wiley & Sons.

Hobbs, R. J., Higgs, E., & Harris, J. A. (2009). Novel ecosystems: Implications for conservation and restoration. *Trends in Ecology & Evolution, 24*(11), 599–605. https://doi.org/10.1016/j.tree.2009.05.012

Holling, C. S. (1978). *Adaptive environmental management and assessment.* Chichester, UK: John Wiley & Sons.

Holmes, H. (1922, June 17). Department of Commerce Blank Book No. 27. Harlan Holmes Papers, Acc. 2614-001, Special Collections, University of Washington Libraries.

Holmes, H. (1923, June 12). Department of Commerce Bureau of Fisheries Blank Book, June 1923. Harlan Holmes Papers, Acc. 2614-001, Special Collections, University of Washington Libraries.

Holmes, H. (1924, October 19). Okanogan Trip 1924 (1). Harlan Holmes Papers, Acc. 2614-001, Special Collections, University of Washington Libraries.

Holmes, H. (1927, August 12). Harlan B. Holmes Field Notes—1927 Columbia River. Harlan Holmes Papers, Acc. 2614-001, Special Collections, University of Washington Libraries.

Holmes, H. (1952). Loss of salmon fingerlings in passing Bonneville Dam as determined by marking experiments. Unpublished manuscript. US Bureau of Commercial Fisheries Report to US Army Corps of Engineers, North Pacific Division, Portland, OR. Available from US Fish and Wildlife Service, Vancouver, WA.

Intergovernmental Panel on Climate Change (IPCC). (2007). *IPCC fourth assessment report: Climate change 2007.* https://www.ipcc.ch/assessment-report/ar4/

Jackson, S. T., & Hobbs, R. J. (2009). Ecological restoration in the light of ecological history. *Science, 325*(5940), 567–569. https://doi.org/10.1126/science.1172977

Jasanoff, S. (2004). *States of knowledge: The co-production of science and social order.* London: Routledge.

Jasanoff, S. (2005). *Designs on nature: Science and democracy in Europe and the United States.* Princeton, NJ: Princeton University Press.

Jasanoff, S. (2015). Future imperfect: Science, technology, and the imaginations of modernity. In S. Jasanoff & S.-H. Kim (Eds.), *Dreamscapes of modernity.* Chicago: University of Chicago Press.

Jasanoff, S., & Kim, S.-H. (2015). *Dreamscapes of modernity: Sociotechnical imaginaries and the fabrication of power.* Chicago: University of Chicago Press.

Jasanoff, S., & Wynne, B. (1998). Science and decision making. In S. Ryner & E. L. Malone (Eds.), *Human choice and climate change* (vol. 1, pp. 1–87). Columbus, OH: Battelle Press.

Johnson, R., Chambers, D., Raghuram, P., & Tincknell, E. (2004). *The practice of cultural studies*. London: Sage Publications.

Jordan, W. R., III, Gilpin, M. E., & Aber, J. D. (1987). Restoration ecology: Ecological restoration as a technique for basic research. In *Restoration ecology: A synthetic approach to ecological research*. Cambridge, UK: University of Cambridge Press.

Kao, S.-C., Sale, M. J., Ashfaq, M., Uria Martinez, R., Kaiser, D. P., Wei, Y., & Diffenbaugh, N. S. (2015). Projecting changes in annual hydropower generation using regional runoff data: An assessment of the United States federal hydropower plants. *Energy, 80*, 239–250. https://doi.org/10.1016/j.energy.2014.11.066

Karasti, H., Baker, K. S., & Millerand, F. (2010). Infrastructure time: Long-term matters in collaborative development. *Computer Supported Cooperative Work (CSCW), 19*(3–4), 377–415. https://doi.org/10.1007/s10606-010-9113-z

Katz, E. (1992). The big lie: Human restoration of nature. *Research in Philosophy and Technology, 12*, 231–242.

Katz, S. L., Barnas, K., Hicks, R., Cowen, J., & Jenkinson, R. (2007). Freshwater habitat eestoration actions in the Pacific Northwest: A decade's investment in habitat improvement. *Restoration Ecology, 15*(3), 494–505. https://doi.org/10.1111/j.1526-100X.2007.00245.x

Keith, D. A., Martin, T. G., McDonald-Madden, E., & Walters, C. (2011). Uncertainty and adaptive management for biodiversity conservation. *Biological Conservation, 144*(4), 1175–1178. https://doi.org/10.1016/j.biocon.2010.11.022

Keller, E. A., & Swanson, F. J. (1979). Effects of large organic material on channel form and fluvial processes. *Earth Surface Processes, 4*, 361–380.

Kingsland, S. E. (2005). *The evolution of American ecology, 1890–2000*. Baltimore, MD: John Hopkins University Press.

Knorr-Cetina, K. (1999). *Epistemic cultures: How the sciences make knowledge*. Cambridge, MA: Harvard University Press.

Kohler, R. E. (2002). *Landscapes and labscapes: Exploring the lab-field border in biology*. Chicago: University of Chicago Press.

Kondolf, G. M., Boulton, A. J., O'Daniel, S., Poole, G. C., Rahel, F. J., Stanley, E. H., . . . & Nakamura, K. (2006). Process-based ecological river restoration: Visualizing three-dimensional connectivity and dynamic vectors to recover lost linkages. *Ecology and Society, 11*(5).

Koontz, T. M., Gupta, D., Mudliar, P., & Ranjan, P. (2015). Adaptive institutions in social-ecological systems governance: A synthesis framework. *Environmental Science & Policy, 53*, 139–151. https://doi.org/10.1016/j.envsci.2015.01.003

Kopf, R. K., Finlayson, C. M., Humphries, N. C., & Hladyz, S. (2015). Anthropocene baselines: Assessing change and managing biodiversity in human-dominated aquatic ecosystems. *BioScience, 65*(8), 798–811.

Kuhn, T. S. (1962). *The structure of scientific revolutions* (2nd ed.). Chicago: University of Chicago Press.

Landeen, D., & Pinkham, A. (1999). *Salmon and his people: A Nez Perce nature guide*. Lewiston, ID: Confluence Press.

Langston, N. (2003). *Where land and water meet: A western landscape transformed.* Seattle: University of Washington Press.
Latour, B. (1999). *Pandora's hope: Essays on the reality of science studies.* Cambridge, MA: Harvard University Press.
Latour, B., & Woolgar, S. (1979). *Laboratory life: The social construction of scientific facts.* Los Angeles: Sage Publications.
Lave, R. (2012). *Fields and streams: Stream restoration, neoliberalism, and the future of environmental science.* Athens: University of Georgia Press.
Lave, R., Doyle, M., & Robertson, M. (2010). Privatizing stream restoration in the US. *Social Studies of Science, 40*(5), 677–703. https://doi.org/10.1177/0306312710379671
Law, J. (1987). Power, action, and belief: A new sociology of knowledge? In *Sociological Review Monograph* (vol. 32). London: Routledge.
Lawler, J. J. (2009). Climate change adaptation strategies for resource management and conservation planning. *Annals of the New York Academy of Sciences, 1162,* 79–98.
Lee, K. N. (1993). *Compass and gyroscope: Integrating science and politics for the environment.* Washington, DC: Island Press.
Leopold, A. (1949). *A sand county almanac: And sketches here and there.* Oxford: Oxford University Press.
Light, A., & Higgs, E. S. (1996). The politics of ecological restoration. *Environmental Ethics, 18,* 21.
Light, A., Thompson, A., & Higgs, E. S. (2013). Valuing novel ecosystems. In R. J. Hobbs, E. S. Higgs, & C. M. Hall (Eds.), *Novel ecosystems* (pp. 257–268). Chichester, UK: John Wiley & Sons. https://doi.org/10.1002/9781118354186.ch31
MacDonald, L. H., Smart, A. W., & Wissmar, R. C. (1991). *Monitoring guidelines to evaluate effects of forestry activities on streams in the Pacific Northwest and Alaska.* Seattle: US Environmental Protection Agency, Region 10, NPS Section.
Mantua, N., Tohver, I., & Hamlet, A. (2010). Climate change impacts on streamflow extremes and summertime stream temperature and their possible consequences for freshwater salmon habitat in Washington State. *Climatic Change, 102*(1–2), 187–223. https://doi.org/10.1007/s10584-010-9845-2
Mao, Y. B., Nijssen, B., & Lettenmaier, D. P. (2015). Is climate change implicated in the 2013–2014 California drought? A hydrologic perspective. *Geophysical Research Letters, 42,* 2805–2813.
McDonald, T. (2013). Adapting restoration and management to climate change. *Ecological Management & Restoration, 14*(3), 158–158. https://doi.org/10.1111/emr.12069
McLain, R. J., & Lee, R. G. (1996). Adaptive management: Promises and pitfalls. *Environmental Management, 20*(4), 437–448.
Mitchell, T. (2014). Introduction: Life of infrastructure. *Comparative Studies of South Asia, Africa and the Middle East, 34*(4), 437–439.
Mote, P. W., Parson, E. A., Hamlet, A. F., Keeton, W. S., Lettenmaier, D., Mantua, N., . . . & Amy K. Snover. (2003). Preparing for climatic change: The water, salmon, and forests of the Pacific Northwest. *Climatic Change, 61,* 45–88.

Mote, P., Rupp, D. E., Li, S., Sharp, D. J., Otto, F., Uhe, P. F., . . . & Allen, M. R. (2016). Perspectives on the causes of exceptionally low 2015 snowpack in the western United States. *Geophysical Research Letters, 43*(20), 10980–10988.

Naiman, R. J., Beechie, T. J., & Benda, L. E. (1992). Fundamental elements of ecologically healthy watersheds in the Pacific Northwest coastal ecoregion. In R. J. Naiman (Ed.), *Watershed management: Balancing sustainability and environmental change* (pp. 127–188). New York: Springer-Verlag.

Naiman, R., Bilby, R. E., & Kantor, S. (1998). *River ecology and management: Lessons from the Pacific coastal ecoregion.* New York: Springer.

National Interagency Fire Center (NIFC). (2017). *Total wildland fires and acres (1960–2017).* Retrieved from https://www.nifc.gov

Nelson, D. R., Adger, W. N., & Brown, K. (2007). Adaptation to environmental change: Contributions of a resilience framework. *Annual Review of Environment and Resources, 32*(1), 395–419. https://doi.org/10.1146/annurev.energy.32.051807.090348

Nesbitt, J. T., & Weiner, D. (2001). Conflicting environmental imaginaries and the politics of nature in Central Appalachia. *Geoforum, 32,* 333–349. https://doi.org/10.1016/S0016-7185(00)00047-6

Nilsson, C., & Aradóttir, Á. L. (2013). Ecological and social aspects of ecological restoration: New challenges and opportunities for northern regions. *Ecology and Society, 18*(4). https://doi.org/10.5751/ES-06045-180435

NOAA Fisheries. (2015). *2015 Adult sockeye salmon passage report.* https://archive.fisheries.noaa.gov/wcr/publications/hydropower/fcrps/2015_adult_sockeye_salmon_passage_report.pdf

Nolin, A., Sproles, E., & Brown, A. (2012). *Transboundary river governance in the face of uncertainty: The Columbia River Treaty; A project of the universities consortium on Columbia River governance.* Corvallis: Oregon State University Press.

Nowotny, H., Scott, P. B., & Gibbons, M. T. (2013). *Re-thinking science: Knowledge and the public in an age of uncertainty.* Cambridge, UK: Polity Press.

NW Council. (1982). *Columbia River Basin Fish and Wildlife Program.* Retrieved from https://www.nwcouncil.org/fish-and-wildlife/previous-programs/1982-columbia-river-basin-fishwildlife-program

NW Council. (2013, December). *Pocket guide: Fast facts about the Columbia Basin.* Retrieved from https://www.nwcouncil.org/reports/pocket-guide

NW Council. (2015, August 12). *Warm water wreaks havoc on Columbia River fish.* Retrieved from www.nwcouncil.org/news/warm-water-wreaks-havoc-columbia-river-fish

NW Council. (2017). *2016 Columbia River Basin Fish and Wildlife Program costs report.* Retrieved from https://www.nwcouncil.org/reports/2016-columbia-river-basin-fish-and-wildlife-program-costs-report

NWF v. NMFS. (2011). National Wildlife Federation v. National Marine Fisheries Service, 839 F. Supp. 2d 1117 (D. Or. 2011).

NWF v. NMFS. (2016). *National Wildlife Federation v. National Marine Fisheries Service*, 184 F. Supp. 3d 861 (D. Or. 2016).

Olsson, P., Gunderson, L., Carpenter, S., Ryan, P., Lebel, L., Folke, C., & Holling, C. S. (2006). Shooting the rapids: Navigating transitions to adaptive governance of social-ecological systems. *Ecology and Society, 11*(1). https://doi.org/10.5751/ES-01595-110118

Oreskes, N. (2003). The role of quantitative models in science. In J. J. Canham, J. J. Cole, & W. K. Lauenroth (Eds.), *Models in ecosystem science* (pp. 13–31). Princeton, NJ: Princeton University Press.

Ostrom, E. (1990). *Governing the commons: The evolution of institutions for collective action.* Cambridge, UK: Cambridge University Press.

Ostrom, E. (2005). *Understanding institutional diversity.* Hoboken, NJ: Princeton University Press.

Ostrom, E. (2009). A general framework for analyzing sustainability of social-ecological systems. *Science, New Series, 325*(5939), 419–422.

Pacific Fishery Management Council. (2016). *Preseason report I: Stock abundance analysis and environmental assessment; Part 1 for 2016 ocean salmon fishery regulations* (Regulation Identifier No. 0648-BF56). Portland, OR: Pacific Fishery Management Council.

Pahl-Wostl, C. (2009). A conceptual framework for analysing adaptive capacity and multi-level learning processes in resource governance regimes. *Global Environmental Change, 19*(3), 354–365. https://doi.org/10.1016/j.gloenvcha.2009.06.001

Paulsen, C. M., & Fisher, T. R. (2005). Do habitat actions affect juvenile survival? An information-theoretic approach applied to endangered Snake River Chinook salmon. *Transactions of the American Fisheries Society, 134*, 68–85.

Pauly, D. (1995). Anecdotes and shifting baselines syndrome of fisheries. *Trends in Ecology & Evolution, 10*, 430.

Pearson, M. L. (2012). The river people and the importance of salmon. In Barbara A. Cosens (Ed.), *The Columbia River Treaty revisited: Transboundary governance in the face of uncertainty.* Corvallis: Oregon State University Press.

Peet, R., & Watts, M. (2002). *Liberation ecologies: Environment, development and social movements.* London: Routledge.

Pickering, A. (1995). *The mangle of practice.* Chicago: University of Chicago Press.

Pollock, M. M., Beechie, T. J., Wheaton, J. M., Jordan, C. E., Bouwes, N., Weber, N., & Volk, C. (2014). Using beaver dams to restore incised stream ecosystems. *BioScience, 64*(4), 279–290. https://doi.org/10.1093/biosci/biu036

Pollock, M. M., Weber, N., & Lewallen, G. (2015, June). Beaver dam analogues (BDAs). In M. M. Pollock, G. Lewallen, K. Woodruff, C. E. Jordan, & J. M. Castro (Eds.), *The beaver restoration guidebook: Working with beaver to restore streams, wetlands, and floodplains* (version 1.0). Portland, OR: US Fish and Wildlife Service. https://www.fws.gov/oregonfwo/toolsforlandowners/riverscience/documents/brg%20v.1.0%20final%20reduced.pdf

Pollock, M. M., Lewallen, G., Woodruff, K., Jordan, C. E., & Castro, J. M. (Eds.). (2015a, June). *The beaver restoration guidebook: Working with beaver to restore streams, wetlands, and floodplains* (version 1.0). Portland, OR: US Fish and Wildlife Service. https://www.fws.gov/oregonfwo/toolsforlandowners/riverscience/documents/brg%20v.1.0%20final%20reduced.pdf

Pollock, M. M., Lewallen, G., Woodruff, K., Jordan, C. E., & Castro, J. M. (Eds.). (2015b, July). *The beaver restoration guidebook: Working with beaver to restore streams, wetlands, and floodplains* (version 1.02). Portland, OR: US Fish and Wildlife Service. https://www.fws.gov/oregonfwo/Documents/BeaverRestGBv.1.02.pdf

Puls, A., Dunn, K. A., & Hudson, B. G. (2014). *Evaluation and prioritization of stream habitat monitoring in the Lower Columbia Salmon and Steelhead Recovery Domain as related to the habitat monitoring needs of ESA recovery plans* (PNAMP Series 2014-003). https://pubs.er.usgs.gov/publication/70191616

Rheinberger, H. J. (2010). *An epistemology of the concrete: Twentieth-century histories of life*. Durham, NC: Duke University Press.

Ribes, D. (2014). Ethnography of scaling or, how to fit a national research infrastructure in the room. In *Proceedings of the 17th ACM conference on computer supported cooperative work and social computing—CSCW '14* (pp. 158–170). Baltimore, MD: ACM Press. https://doi.org/10.1145/2531602.2531624

Ribes, D. (2014). The kernel of a research infrastructure. In *Proceedings of the 17th ACM conference on computer supported cooperative work and social computing—CSCW '14* (pp. 574–587). Baltimore, MD: ACM Press. https://doi.org/10.1145/2531602.2531700

Ribes, D. (2017). Notes on the concept of data interoperability: Cases from an ecology of AIDS research infrastructures. In *Proceedings of the 2017 ACM conference on computer supported cooperative work and social computing—CSCW '17* (pp. 1514–1526). Baltimore, MD: ACM Press. https://doi.org/10.1145/2998181.2998344

Ribes, D., & Finholt, T. (2009). The long now of technology infrastructure: Articulating tensions in development. *Journal of the Association for Information Systems, 10*(5). http://aisel.aisnet.org/jais/vol10/iss5/5

Ribes, D., & Polk, J. (2014). Flexibility relative to what? Change to research infrastructure. *Journal of the Association for Information Systems, 15*(5). https://aisel.aisnet.org/jais/vol15/iss5/1

Ribes, D., & Polk, J. B. (2015). Organizing for ontological change: The kernel of an AIDS research infrastructure. *Social Studies of Science, 45*(2), 214–241.

Rieman, B. E., & Isaak, D. J. (2010). *Climate change, aquatic ecosystems, and fishes in the Rocky Mountain West: Implications and alternatives for management* (No. RMRS-GTR-250). Ft. Collins, CO: US Department of Agriculture, Forest Service, Rocky Mountain Research Station. https://doi.org/10.2737/RMRS-GTR-250

Rieman, B. E., Smith, C. L., Naiman, R. J., Ruggerone, G. T., Wood, C. C., Huntly, N., . . . & Smouse, P. (2015). A comprehensive approach for habitat restoration in the Columbia Basin. *Fisheries, 40*(3), 124–135. https://doi.org/10.1080/03632415.2015.1007205

Roberts, C. D., Palmer, M. D., McNeall, D., & Collins, M. (2015). Quantifying the likelihood of a continued hiatus in global warming. *Nature Climate Change, 5*(4), 337–342. https://doi.org/10.1038/nclimate2531

Robinson, J. (2015). Being undisciplined: Transgressions and intersections in academia and beyond. *Futures, 40*(1), 70–86.

Roni, P. (Ed.). (2005). *Monitoring stream and watershed restoration.* Bethesda, MD: American Fisheries Society.

Rosgen, D. L. (1994). A classification of natural rivers. *Catena, 22,* 169–199.

Ruhl, J. B., & Fischmann, R. L. (2011). Adaptive management in the courts. *Minnesota Law Review, 95,* 424–484.

Ruhl, J. B., & Salzman, J. (2011) Gaming the past: The theory and practice of historic baselines in the administrative state. *Vanderbilt Law Review, 64,* 1–57.

Salo, E. O., & Cundy, T. W. (Eds.). (1987). *Streamside management: Forestry and fishery interactions.* Seattle: College of Forest Resources, University of Washington.

Schumm, S. A., Harvey, M. D., & Watson, C. C. (1984). *Incised channels: Morphology, dynamics, and control.* Littleton, CO: Water Resources Publications.

Scott, J. C. (1999). *Seeing like a state: How certain schemes to improve the human condition have failed.* New Haven, CT: Yale University Press.

Seastedt, T. R., Hobbs, R. J., & Suding, K. N. (2008). Management of novel ecosystems: Are novel approaches required? *Frontiers in Ecology and the Environment, 6*(10), 547–553.

Society for Ecological Restoration International Science & Policy Working Group. (2004). *The SER International primer on ecological restoration.* Retrieved from www.ser.org

Stanford, J. A., Frissell, C. A., & Coutant, C. C. (2006). The status of freshwater habitats. In *Return to the river: Restoring salmon to the Columbia River.* Burlington, VT: Elsevier Academic Press.

Star, S. L., & Bowker, G. (2010). How to infrastructure. In L. A. Lievrouw & S. Livingstone (Eds.), *Handbook of new media: Social shaping and social consequences of ICTs.* London: Sage Publications.

Star, S. L., & Ruhleder, K. (1994). Steps towards an ecology of infrastructure: Complex problems in design and access for large-scale collaborative systems. In *Proceedings of the conference on computer supported cooperative work.* Chapel Hill, NC: ACM Press.

Starzomski, B. M. (2013). Novel ecosystems and climate change. In R. J. Hobbs, E. S. Higgs, & C. M. Hall (Eds.), *Novel ecosystems* (pp. 88–101). Chichester, UK: John Wiley & Sons.

Steinhardt, S. B., & Jackson, S. J. (2015). Anticipation work: Cultivating vision in collective practice. In *Proceedings of the 18th ACM conference on computer supported cooperative work and social computing* (pp. 443–453). New York: ACM Press. https://doi.org/10.1145/2675133.2675298

Suding, K., & Leger, E. (2012). Shifting baselines: Dynamics of evolution and community change in a changing world. In J. van Andel & J. Aronson (Eds.), *Restoration ecology: The new frontier* (2nd ed., pp. 281–292). Chichester, UK: John Wiley & Sons. https://doi.org/10.1002/9781118223130.ch21

Tadaki, M., & Sinner, J. (2014). Measure, model, optimise: Understanding reductionist concepts of value in freshwater governance. *Geoforum, 51*, 140–151.

Taylor, J. E. (1999). *Making salmon: An environmental history of the Northwest fisheries crisis.* Seattle: University of Washington Press.

Taylor, P. J. (2005). *Unruly complexity.* Chicago: University of Chicago Press.

Thompson, D. M. (2006). Did the pre-1980 use of in-stream structures improve streams? A reanalysis of historical data. *Ecological Applications, 16*, 784–796.

Thompson, D. M., & Stull, G. N. (2002). The development and historic use of habitat structures in channel restoration in the United States: The grand experiment in fisheries management. *Geographie physique et Quaternaire, 56*, 45–60.

Ureta, S. (2017). Baselining pollution: Producing "natural soil" for an environmental risk assessment exercise in Chile. *Journal of Environmental Policy and Planning, 20*(7), 1–14.

US Entity Treaty Review. (2013). *US Entity regional recommendation for the future of the Columbia River Treaty after 2024.* https://www.bpa.gov/Projects/Initiatives/crt/CRT-Regional-Recommendation-eFINAL.pdf

US Forest Service (USFS). (2018). *Challis-Yankee Fork Ranger District—Sunbeam Dam.* Retrieved from https://www.fs.usda.gov/detail/scnf/about-forest/districts/?cid=fsbdev3_029697

van Diggelen, R., Grootjans, A. P., & Harris, J. A. (2001). Ecological restoration: State of the art or state of the science? *Restoration Ecology, 9*(2), 115–118. https://doi.org/10.1046/j.1526-100X.2001.009002115.x

van Hemert, M. (2013). Taming Indeterminacy: The co-production of biodiversity restoration, flood protection and biophysical modelling of rivers and coastal environments. *Science, Technology and Society, 18*(1), 75–92. https://doi.org/10.1177/0971721813484376

Vano, J. A., Nijssen, B., & Lettenmaier, D. P. (2015). Seasonal hydrologic responses to climate change in the Pacific Northwest. *Water Resources Research, 51*(4), 1959–1976.

Volkman, J. M., & McConnaha, W. E. (1993). Through a glass, darkly: Columbia River salmon, the Endangered Species Act, and adaptive management. *Environmental Law, 23*(4), 1249–1272.

Walter, A. I., Helgenberger, S., Wiek, A., & Scholz, R. W. (2007). Measuring societal effects of transdisciplinary research projects: Design and application of an evaluation method. *Evaluation and Program Planning, 30*(4), 325–338.

Walter, R. C., & Merritts, D. (2008). Natural streams and the legacy of water-powered mills. *Science, 319*(5861), 299–304.

Walters, C. J., & Holling, C. S. (1990). Large-scale management experiments and learning by doing. *Ecology, 71*(6), 2060–2068. https://doi.org/10.2307/1938620

Walters, C., & Hilborn, R. (1976). Adaptive control of fishing systems. *Journal of the Fisheries Research Board of Canada, 33*, 145–159.

Walters, C. J., & Riddell, B. (1986). Multiple objectives in salmon management: The Chinook sport fishery in the Strait of Georgia, British Columbia. *Northwest Environmental Journal 2*, 1–15.

Weber, M. (1946). *Max Weber: Essays in sociology*. Oxford: Oxford University Press.

West, J. M., Julius, S. H., Kareiva, P., Enquist, C., Lawler, J. J., Petersen, B., . . . & Shaw, M. R. (2009). U.S. natural resources and climate change: Concepts and approaches for management adaptation. *Environmental Management, 44*(6), 1001. https://doi.org/10.1007/s00267-009-9345-1

Wheaton, J. M., Bennett, S. N., Bouwes, N., Maestas, J. D., & Shahverdian, S. M. (Eds.). (2019). *Low-tech process-based restoration of riverscapes: Design manual* (version 1.0). Logan: Utah State University Restoration Consortium. http://lowtechpbr.restoration.usu.edu/manual

White, R. (1995). *The organic machine: The remaking of the Columbia River*. New York: Hill and Wang.

White, R. J. (1996). Growth and development of North American stream habitat management for fish. *Canadian Journal of Fisheries and Aquatic Science, 53*(1), 342–363.

Whitehead, A. N. (1927–1928). *Process and reality*. D. R. Griffin & D. W. Sherburne (Eds.). New York: Free Press. (Originally published 1978)

Wilkinson, C. F. (1992). *Crossing the next meridian: Land, water, and the future of the West*. Washington, DC: Island Press.

Williams, J. W., & Jackson, S. T. (2007). Novel climates, no-analog communities, and ecological surprises. *Frontiers in Ecology and the Environment, 5*(9). http://www.jstor.org/stable/20440743

Williams, R. N. (Ed.). (2006). *Return to the river: Restoring salmon to the Columbia River*. Burlington, VT: Elsevier Academic Press.

Woelfle-Erskine, C. (2017). The watershed body: Transgressing frontiers in riverine sciences, planning stochastic multispecies worlds. *Catalyst: Feminism, Theory, Technoscience, 3*(2), 1–30.

Wohl, E., Lane, S. N., & Wilcox, A. C. (2015). The science and practice of river restoration. *Water Resources Research, 51*, 5974–5997.

Wohl, E., Angermeier, P. L., Bledsoe, B., Kondolf, G. M., MacDonnell, L., Merritt, D. M., . . . & Tarboton, D. (2005). River restoration. *Water Resources Research, 41*, W10301. https://doi.org/10.1029/2005WR003985

Woodruff, K. (2016). *Methow Beaver Project accomplishments 2015*. https://www.seattle.gov/light/Environment/WildlifeGrant/Projects/Woodruff%202016%20Methow%20Beaver%20Project%20Accomplishments.pdf

Worster, D. (1987). *Nature's economy: A history of ecological ideas* (2nd ed.). Cambridge, UK: Cambridge University Press.

Zabel, R., Cooney, T., & Jordan, C. (2013). Introduction. In R. Zabel et al. (Eds), *Life-cycle models of salmonid populations in the interior Columbia River Basin*. Seattle: NOAA Northwest Fisheries Science Center.

Zedler, J. B., Doherty, J. M., & Miller, N. A. (2012). Shifting restoration policy to address landscape change, novel ecosystems, and monitoring. *Ecology and Society*, *17*(4). https://doi.org/10.5751/ES-05197-170436

Zedler, J. B., & Callaway, J. C. (2003). Adaptive restoration: A strategic approach for integrating research into restoration projects. In D. J. Rapport, W. L. Lasley, D. E. Rolston, N. O. Nielsen, C. O. Qualset, & A. B. Damania (Eds.), *Managing for Healthy Ecosystems* (pp. 167–174). Milton Park, UK: Routledge.

INDEX

A

acclimation
  adaptation vs., 132
  future of, 123–25, 144–45
  meaning of, 131–32
  monitoring programs, 126–33
  by restorationists, 15
  term use, 170
accommodation, meaning of, 30, 117
Action Effectiveness Monitoring of Tributary Habitat Improvement (AEM), 126
adaptation
  acclimation vs., 132
  adaptive management vs., 132
  defined, 12
  in knowledge, 29–30
  by restorationists, 169–70
  in science, 12–13, 28, 30–34, 84
  in socio-ecological systems, 135–36
adaptive capacity, 18–19, 174
adaptive cycle of panarchy, 117
adaptive epistemologies
  co-production, 174–78
  defined, 36
  imaginaries, environmental and socio-technical, 178–83
  making strategies visible, 171–74
adaptive knowledge infrastructures, 125, 159–61, 169–70
adaptive learning, modeling for, 147

adaptive management
  adaptation vs., 132
  adaptive knowledge supporting, 29–30
  defined, 51
  emergence concept in, 117–20
  failures of, 51–52
  implementation of, 27–28
  literature on, 27
  monitoring in, 125
  policy making, incorporating into, 51–53
  pragmatic, 181–82
  in salmon restoration, 20–21
  in socio-ecological systems, 18–22
  success of, determining, 20
adaptive management implementation plan (AMIP), 51–52
agricultural development, 44–45, 56, 89
air temperatures, climate change effects on, 96–98, 100, 169
Allison, Stuart, 66
Anthropocene baselines, 96
Anticipating Future Environments (Hirsch)
  methods, 26–27
  situating the study, 24–26
  transdisciplinarity in, 22–25
anticipation, defined, 147, 159, 163, 170
anticipation in models
  indeterminacy in, 156–59
  life-cycle, 154–56
  as metaphors, 161–65

205

anticipation in models (*continued*)
    stream temperature and water-quality, 150–54
Army Corps of Engineers (Corps), 39, 50, 54, 57–58, 75

B

Bahls, Peter, 89–90, 92, 94, 97–98, 101–2, 167
baselines
    acclimation of, 133
    Anthropocene, 96
    environmental, 94–96
    ESA requirement for, 95–96
    establishing, 148
    metrics, changes in, 138–44
Bayer, Jennifer, 134, 139, 142
Bear Valley, 57
beaver, 93, 109–14
beaver dam analogs (BDAs), 108–9, 111, 115–16, 118–20
beaver dams, 109–12, 115–16
beaver restoration, 104–5, 109–16, 119, 140
The Beaver Restoration Guidebook (Pollock et al.), 114
Beechie, Tim, 75–76, 80, 157
Beyeler, Merrill, 166–69
Biogenic dams, 110
biological opinions (BiOps), 50–52, 139, 155
Bitterroot Mountain, 166
the blob, 97–98, 129
Boise River, 56
Bonneville Dam, 46, 55, 58–59, 72, 179
Bonneville Power Administration (BPA)
    habitat monitoring, funding, 126, 130
    habitat monitoring programs, 124
    habitat restoration, funding, 59, 79, 81, 130
    Lemhi restoration project and the, 168
    NOAA, dam consultations with the, 50
    policy documents, analysis of, 27

Bowker, Geoffrey, 140–42, 144
Bradshaw, Anthony, 81, 123
Braverman, Irus, 100–101
Bureau of Fisheries, 38–39, 42, 55
Bureau of Reclamation (BoR), 44, 50, 71, 112
Bush (George W.) administration, 51

C

Cabin, Robert, 83
Canada, 25, 48
Castro, Janine, 118, 122
Celilo Falls, 46
Center for Streamside Studies, 80
Channel Evolution Model, 62
Civilian Conservation Corps (CCC), 75
Clark Fork River, 43
Clean Water Act, 78
Cle Elum, WA, 55
climate adaptation, defined, 12
climate change
    beaver restoration and, 116
    belief in, 135–37
    effects of in the Columbia River Basin, 5–6, 8–10, 96–100, 108, 151, 166–67, 169
    mitigation and adaptation policy, 21, 98–99
    modeling, 147, 151–55, 158
    novel solutions to, 119
    preparing for, 172–74
    restorationists use of data on, 111–12
    Trump administration on, 24–25
    uncertainty introduced with, 5–10, 15–16, 119
    unpredictability of, 121
cold-water refugia, 167
collective empiricism, 175
Columbia Basin tribes, 53
Columbia Basin Water Transactions Program, 168
Columbia Habitat Monitoring Program (CHaMP), 126, 130

Columbia River Basin
    change in the, 28, 39, 110
    dam development era, 44–47, 71
    development of the, 28, 43–44, 110
    diversity of the, 43
    historical survey of the, 38
    length of the, 43
    map of the, 7
    "organic machine" of the, 45, 47, 73, 172, 180
    restoration of the, 31, 58–61
    spatial and jurisdictional complexities, 48–49, 70
    tribes and cultures in the, 43
    vastness of the, 3–4, 43
Columbia River Fish and Wildlife Program, 79
Columbia River Inter-Tribal Fish Commission (CRITFC), 49, 80, 84
Columbia River Treaty, 47–48, 78
complexity, spatial, 70
controllability, partial, 149
Convention for Biological Diversity (CBD), 98
Coordinated Assessment project, 133
Correll, L. W., 58
creativity, science and, 178
cultures
    technoscientific, 109

D

Dabob Bay, 89
Dalles Dam, 46
dam development era, 44–47, 71
dams
    damage done by, mitigating, 50–51, 73, 77–79
    FCRPS management of, 50
    migration, effect on, 7, 48, 72
    salmonids survival, effect on, 50–51
    salmon survival, effects on, 4, 48, 55–56, 72–73, 77–78
    sociotechnical nature of, 45
    See also hydropower dams; specific dams
Daston, Lorraine, 32–33, 121, 136, 175
decision making, embracing unknowns in, 21–22
DiLuccia, Jeff, 166–69
drones, use in restoration, 134
drought, 6, 169. See also snowmelt
Dworshak Dam, 47

E

Ecological Restoration (journal), 67
ecological science, 68–69, 71–72
Ecological Society of America, 67
ecosystem equilibrium models, 99
ecosystems
    hybrid, 16, 99
    no-analog, 21
    novel, 16, 99
    restoration, 99–100
emergence concept, 35, 107–9, 115–21, 170
Endangered Species Act (ESA)
    adaptive capacity, 19
    co-production and the, 58–61
    drivers of, 84–85
    fish counting focus in, 138–39
    goals, restoration and recovery, 96, 138–40
    knowledge production, 11
    legitimacy of the, 19
    listings, 47, 50–52, 60, 80, 177
    management interventions driving, 78
    monitoring programs, 80–81
    precautionary principle of the, 33
    purpose of, 53
    requirements of the: anticipation component in, 163; baselines, 95–96; BiOps, 79; meeting the, 145, 171; metrics, 69, 141; for mitigation, 50; for recovery, 60; for restoration, 49–50, 123

INDEX 207

engineering-based restoration
  dominant paradigm of, 112–13, 115
  process-based vs., 73–77, 107–9, 118–19, 121–22
environmental imaginaries
  components of, 68–69
  co-producing, 174–78
  future research, suggestions for, 178–83
  making visible, 171–74
  purpose in, 101
  river futures and, 166–71
environmental management, co-production in, 58–61, 156
environmental variables, measurability, 135–36
environmental variation, 149
epistemic community of restoration ecologists
  adaptation within the, 14–17, 169–70
  challenges for, 86–87
  characteristics of, 14–15, 107
  work of the, 4
epistemic culture, 28–33, 108–9, 117–19
epistemic virtues, 32–33, 118–19, 121
ethnographic methods, use of, 24–26, 31, 35, 100, 173

F

Federal Columbia River Power System (FCRPS), 50
Fish Accords, 84
Fish and Wildlife Program, 81
Fish and Wildlife Service, 39, 47, 50, 55
fish bypass technology, 55–56, 77–78
fish counting, 129–30, 138–39
fishing industry, 39, 153
fish marking project of Harlan Holmes, 39–42
fish passage projects, 73
fish passage technology, 55, 77–78
flights of speculative imagination, 177–78
floodplain connectivity, 74, 167–68

fly-fishing industry, 153
forestry industry, 79–80
Forest Service (USFS)
  beaver restoration project and the, 105, 114
  Great Basin Experiment Station, 66
  NorWeST Stream Temp project, 150–51, 153–54, 159, 167
  streambank stabilization, 77
forward-looking infrared (FLIR) technology, 134
Furfey, Rosemary, 60
futures
  aligning practices and knowledge infrastructures toward, 148
  anticipatory, 156, 159–60, 161–66
  material, 11
  modeling, 35
  no-analog, 96, 99
futures imaginaries, 11, 166–71. *See also* environmental imaginaries

G

Galison, Peter, 32–33, 121, 136, 175
Gallagher, Sean, 89–94, 100–103, 167
Gore, Al, 151
governance
  adaptive, 18–19
  anticipatory, 172
Grand Coulee Dam, 45, 46
Grande Ronde Model Watershed, 111
Great Basin Experiment Station (USFS), 66

H

habitat loss, major factors in, 54
habitat restoration
  adaptation in, 17, 169–70
  adaptive management approach to, 20–21
  baseline-setting practice, 95

beaver-driven, 109–16
climate change and, 5–10, 15–16, 28, 96–103
collective empiricism vs. competitive models for, 176
coordinated, emergence of, 77–81
crises and paradigm shifts in, 29
criticality of, 6, 54, 100
cultures of, 31
defined, 67
development of, 69–72
drivers of, 84–85
drones use in, 134
ecosystem-based approach to, 81, 99–100
ESA mandate driving, 123
field-centered, 148
funding, 48–49, 51, 59, 73, 129–31, 135
global prioritization of, 98
historically, 34, 38–42, 54–60, 69–73, 108
intelligent tinkering model of, 83
life-cycle models in, 154–56
meaning of, 15
modeling, 148, 150–54, 156
monitoring, 125–26, 130
new paradigms in, 16
no-analog future of, 16–17
novel solutions to, 119
possibility of, 6, 16–17, 101–2, 182–83
reconceptualizing, 99–100
requirements for, 167–68
research study, 24, 27
restoration ecology vs., 67
science of, 4, 14–15
science-practice divide, 83–84
spatial and jurisdictional complexities, 48–49
stakeholders in, 4–6, 49, 84
success and failure in, measuring, 94–95, 122–23, 135–44
See also restoration ecology
Hacking, Ian, 29
hatchery science, 38–41, 54, 70, 73, 129–30

Hells Canyon Dam, 47
Higgs, Eric, 82–83
high-modernism, 46
Hilborn, R., 19
Hobbs, R. J., 100
Holmes, Harlan, 34, 38–42, 54–60, 77–78, 85–86, 129–30, 152, 168, 178–79
Hood Canal, 98
hybrid ecosystems, 16, 99
hydropower
    climate change effects on, 9
    dividing the Columbia River Basin, effects of, 72
    historically, 46
    Native American tribes, effect on, 53
    spatial and jurisdictional complexities, 48
    2008 BiOp failure, 155
hydropower dams
    impact of, 43, 46
    irrigation accompanying, 44–45
    migration, effect on, 7, 72
    number of, 46
    See also dams; *specific dams*

I

imaginaries. *See* environmental imaginaries; sociotechnical imaginaries
Independent Scientific Review Panel (NW Council), 81, 157
indeterminacy, epistemic, 36
Indigenous peoples
    First Food of, 43, 47, 49, 53
    fishing rights, 43, 78
    future of the, 173–74
    power of, 78
    restoration, demand for, 80
    salmon run declines, effects on, 72
    treaty rights, 47–48, 53, 173
institutions
    adaptive, 13, 171, 176
    legal, 19, 59, 163
    organizations and, 174

INDEX 209

in-stream restoration, 74
Integrated Status and Effectiveness Monitoring Program (ISEMP), 126, 130
Integrated Status and Trend Monitoring Project (ISTM), 133
intensively monitored watersheds (IMW), 126, 130–31
interdisciplinary work, defined, 23
irrigation projects, 44–45, 56, 167
Isaak, Dan, 150–53, 160

J

Jasanoff, Sheila, 42, 70
John Day Fisheries Mitigation, 73
justice
    ecological, 72
    environmental, 72, 78

K

Keechelus River Dam, 55
Knorr-Cetina, Karin, 30–31
knowledge infrastructures
    acclimating to climate change, 143–45
    adaptive, 33–34, 125, 159–61, 169–70
    change in, 182
    creating, 140–41, 143–44
    defined, 33–34
    flexible, designing, 172
    with futures orientation, 140
    large-scale, supporting habitat restoration, 34
    standards in, 140, 159
knowledge production
    adaptive change, role in, 17, 170, 174
    change in, 34, 148, 182
    importance of, 11–12
Kuhn, Thomas, 28, 118

K

labor, scientific, 45
Lake Mead, 104
Lake Quinault, 40–41

large woody debris (LWD), 74, 77, 80, 91–93
law
    adaptive, 18–19
    environmental, 19, 84
    law and science, co-production of, 42, 58–61, 174
Lee, Kai, 20, 51
Lemhi Regional Land Trust, 168
Lemhi Valley, 166–69, 177
Leopold, Aldo, 65–66, 75, 83
lidar, 134, 168
life-cycle models, 154–56
logging industry, 79–80
long-now of infrastructure, 163, 165
Lower Columbia River Fisheries Development Program, 73

M

macroinvertebrates, 142–43, 170
Malheur River, 46
mangle of practice, 30
Marsh, George Perkins, 66
material agency, 120
Methow Beaver Project, 105, 112
Methow River, 56, 104–7, 111–12
migration
    of anadromous fish, 167
    assisted, 16
    climate change, effect on, 5–6, 8
    Columbia River Basin map of, 7
    dams effect on, 7, 46, 48, 57–58, 72
    miles crossed during, 4
    monitoring patterns of, 39–42
Migratory Bird Habitat Stamp Act, 70
mining industry, 56, 75
Mitchell Act, 72–73
modeling
    adaptive learning through, 147
    climate change, 147, 151–55, 158
    field-collected data informing, 148, 157
    purposes of, 146–48

temperature, in habitat restoration, 150–53, 156
uncertainty and, 148–50, 156–58
See also anticipation in models
models, types of
  ecosystem-based, 97
  ecosystem equilibrium, 15, 99
  hydrological, 147
  life-cycle, 154–56
  population, 147
  predictive, 148, 158, 162
  single-channel, 63
  spatial statistical, 150
  stream temperature, 149–50
modernity, 36, 171
Molesworth, Jennifer, 111–13, 182–83
monitoring studies, 124–31
Mote, Phil, 151

N

National Environmental Policy Act (NEPA), 78, 84
National Oceanic and Atmospheric Administration Marine Fisheries Service (NOAA Fisheries), 49–53, 60, 80, 84, 124, 155
National Research Council, 81
National Science Foundation (NSF), 59
Native American tribes. *See* Indigenous peoples
Natural Channel Design, 63
natural resource management, 68
natural resource management agencies, 20
The Nature Conservancy, 168
New Deal, 46
Nez Perce Tribe, 43, 78
no-analog ecosystems, 21
no-analog future, 96, 99
Northwest Forest Plan, 79–80, 84
Northwest Power and Conservation Council, (NW Council), 20, 27, 48–49, 75, 84, 124, 127
  BPA partnership, 59, 81
  die-off statistics, 8
  Fish and Wildlife Program, 81
  formation and purpose of, 79, 81
  Independent Scientific Review Panel, 157
Northwest Power and Planning Act, 79
Northwest Watershed Institute (NWI), 89–93, 101
NorWeST Stream Temp project, 150–51, 153–54, 159, 160, 167
novel ecosystems, 16, 99
novelty, emergence of, 178
NWF v. NMFS, 52

O

Obama (Barack) administration, 51
observability, partial, 149
Okanagan River Valley, 55–56
"organic machine" of the Columbia River Basin, 45, 47, 73, 172, 180
organizations, institutions and, 174
Ostrom, Elinor, 13, 117, 135
Othello Sandhill Crane Festival, 3
Oveson, Jeff, 111

P

Pacific Northwest Aquatic Monitoring Partnership (PNAMP), 139–42, 145
panarchy, 117
paradigm shift
Pauly, Daniel, 95
Pess, George, 80
Pickering, Andrew, 30, 120, 162
Plant-A-Thon (NWI), 101
policy and science, co-production of, 96, 123–24, 133, 138–41, 151, 169, 172, 175
post-normal science, 29
Powell, John Wesley, 66
process-based restoration
  critics and supporters of, 111–13
  engineering-based vs., 73–77, 106–9, 118–19, 121–22
  example of, 93

R

Ravetz, Jerome, 29
reasonable and prudent alternatives (RPAs), 50
Redden, James A., 52
Regional Habitat Indicator Project, 142
resilience, in socio-ecological systems, 17, 21
resistance, 30, 117
restoration, monitoring
    adaptations required in, 123
    adaptive management approach to, 21
    baselines, 126
    basin-wide coordinated efforts, 132–33
    challenges in, 124, 126–31
    conflicts and tensions in, 124
    data collection, changes in, 133–34
    focuses of, 124
    funding, 59, 126–27, 129–30, 136
    for the future, 127–33, 144–45
    metrics, changes in, 125–26, 138–44
    modeling and, 147
    observational strategies, 137–38
    trained judgement method of, 135–38
    of watersheds, 48, 80–81, 126–31
restoration cultures, 31, 35, 109
restoration ecologists
    epistemic community of: adaptation within the, 14–17, 169–70; challenges for, 86–87; characteristics of, 14–15, 107; work of the, 4
    future orientation of, 147–48
    strategies for dealing with risk, 119
    term use, 67
    See also specific restorationists
restoration ecology
    challenges to, 86–87
    debates within the field of, 81–84
    origins of, 65–69
    research funding structure, 59
Restoration Ecology (journal), 67, 83
restoration industry, 4
restoration planning, 9–10, 16–17
restoration science
    ESA-centric, 138–39
    history of, 12, 176–77
restoration work
    baselines requirements in, 94–96
    climate change and, 96–102
    drivers of, 84–86
    success of, measuring the, 64, 69–70
Return to the River, 81
riparian restoration, 71, 76–77, 79–80, 105, 167–69
risk, managing, 20, 119
River Restoration Northwest Symposium, 14, 62, 69, 76
rivers
    climate change, effects on, 5–8, 96–98, 104, 111, 115, 167, 169
    floodplain connectivity, 167–68
    restoration of, 73–77, 80–81
Rivers and Harbors Act, 54
Rosgen, Dave, 63
Rosgen system, 63
Rubin, Jude, 90–91, 101

S

salmon, life cycle of, 129
salmon crisis (1960s), 39
salmon extinction, 4, 12
salmonids
    categorization of, 139
    decline, major factors in, 54–56
    life cycle, 47–48
    migration, 47–48, 97–98
    recovery, requirements for, 168–69
salmon populations, declines in
    addressing, preferred methods for, 54
    areas of, 60
    biodiversity, 12
    causes of, 39, 43–44
    climate change and, 5–6, 8–9, 108
    dams effect on, 4, 46–47, 57–58, 72–73

Indigenous peoples, effects on, 72
juvenile, 55
mitigation efforts, 77–79
salmon recovery
    adaptive management approach to, 20–21
    collaboration in working toward, 177
    collective empiricism vs. competitive models for, 175–76
    drivers of, 84–85
    ESA mandates driving, 60
    factors in, 57, 129, 133–34
    intricacy of, 48
    legal requirements for, 139
    modeling, 149
    monitoring, 128–29
    requirements for, 167–68
    success and failure in, measuring, 96, 141–43
    See also habitat restoration
Salmon River, 43, 56–57
Salmon River Sub-Basin, 56
salmon runs, 70, 72
science
    adaptation in, 12–13, 28, 30–34, 84, 117
    anticipatory governance for, 172
    anticipatory strategies for, 159
    change in, 120, 123–24
    creativity and, 178
    crises and paradigm shifts in, 28–29
    emergence in, 117
    giving voice to, 174
    and law, co-production of, 42, 58–61, 174
    norms of, 32
    objectivity in, 32
    and policy, co-production of, 96, 123–24, 133, 138–41, 151, 169, 172, 175
    post-normal mode of operation, 29
    and society, co-production of, 163
    sociotechnical nature of, 45
    stabilities in, 29
    technoscientific changes in, 133

science and technology studies (STS), 27–29, 31, 33–34
science-practice divide, 83–84
science studies, 28–29
scientific cultures, 30, 117–19. *See also* epistemic cultures
scientific practice, 30–36
scientific virtues, 121
shifting baselines syndrome, 95–96
Shingle Creek, 56
Shoshone-Bannock Tribes, 50
Simon, Michael A., 52
situated knowledge, 26
situational analysis, 26
Snake River, 8, 43, 44, 46, 50, 56
snowmelt
    climate change effects on, 5–6, 8, 98, 104–6, 115, 169
    losses in, beavers in mitigating, 111, 116
society and science, co-production of, 163
Society for Ecological Restoration, 67, 76
socio-ecological systems
    adaptive, 18–19, 135, 176–77
    adaptive management of, 19–22, 120
    change theory in, 18
    emergence concept in, 117–20
    environmental variables, measurability of, 135
    framework of, 17–18
    managing, 174
    measurability of variables in, 135–38
    research on, 18
socio-ecological systems theory, 176–77
sociotechnical imaginaries, 11–12, 163, 171, 178–83
spatial complexity, 70
spatial disparity, 72
Spirit of the Salmon: Wy-Kan-Ush-Mi Wa-Kish-Wit, 80
Stage 0 floodplains, river valleys, and streams, 57, 69–70, 110, 168, 170

Star, Susan Leigh, 140–42, 144
Stream Evolution Model, 63
StreamNet Project (PSMFC), 133
stream restoration, 73–77, 80–81, 112, 167–68
streams
    climate change effects on, 5–6, 8, 96–98, 167
    data collection methods, changes in, 134
    dewatering through irrigation, 167
    Stage 0, 69–70, 168
Streamside Management (College of Forest Resources), 79
stream temperature models, 149–50
structural uncertainty, 149
The Structure of Scientific Revolutions (Kuhn), 28
Sunbeam Dam, 56
Supplemental BiOp, 52

T

Tarboo Creek–Dabob Bay watershed, 89–95, 97–98, 101, 167
technoscientific solutions, 109, 172
temperature modeling, 149–53, 156
Teton Dam, 47
"308 Reports" (Corps), 54
transboundary, 48
transdisciplinary, 22–24
"Transplanting Beavers by Airplane and Parachute" (Heter), 110
treaty rights, 49, 53–54, 78, 173, 180–82
Trump, Donald, 24

U

Umatilla Tribe, 78
uncertainty, 20–22, 148–50, 156–58
United Nations Environment Program (UNEP), 98
unruly complexity, 118–19
Upper Snake River, 47, 110
Upstream (NRC), 81
US v. Oregon, 78

W

Warm Springs Dam, 46
Warm Springs Tribe, 78
watershed restoration
    baselines, 89, 103
    beaver's role in, 107, 121
    funding, 127
    monitoring, 48, 80–81, 126–31
    possibility of, 96
    Tarboo Creek, 91–94
watersheds, 5–6, 8, 98
water storage, beaver and, 111, 116
water temperatures, climate change effects on, 5–6, 8, 96–98, 167
Weber, Max, 161
wetlands, 70–71
Wheaton, J. M., 115
White, Richard, 34, 45, 180
Whitehead, Alfred North, 177–78
wildfires, 6, 8, 100, 166, 169
Willamette River, 43
Woodruff, Kent, 104–8, 110, 116
Worster, Donald, 68

Y

Yakima Project, 55
Yakima Tribe, 78
Yakima Valley, 55

Z

Zabel, Rich, 154 55